Peter Wyatt Squire

Methods and formulae used in the preparation of animal and vegetable tissues for microscopical examination

Peter Wyatt Squire

Methods and formulae used in the preparation of animal and vegetable tissues for microscopical examination

ISBN/EAN: 9783337228583

Printed in Europe, USA, Canada, Australia, Japan

Cover: Foto ©berggeist007 / pixelio.de

More available books at **www.hansebooks.com**

METHODS AND FORMULÆ

USED IN THE PREPARATION OF

ANIMAL AND VEGETABLE TISSUES

FOR MICROSCOPICAL EXAMINATION

INCLUDING THE STAINING OF

BACTERIA

BY

PETER WYATT SQUIRE

FELLOW OF THE LINNEAN SOCIETY

LONDON
J. & A. CHURCHILL,
11, NEW BURLINGTON STREET
1892
Copyright Registered.

PREFACE.

THIS little book has grown out of an attempt to frame a collection of trustworthy formulæ most likely to be useful to the practical microscopist. Interspersed with these have been added such practical notes and comments as have suggested themselves during a very extensive course of experimenting on the various materials and processes here described.

The larger text-books do not always agree in the figures given for the same formula. Whenever possible, these have been corrected by reference to the original or most authentic source, but where the formula has been so vague and indefinite as to be little more than a suggestion, a detailed process has been worked out experimentally.

Although intended primarily as a guide in staining processes, the most approved methods of hardening, fixing, embedding, clearing and mounting, are also described. Section-cutting alone has been purposely omitted, as this art can only be learned properly from personal tuition.

Although the best results in staining sections, &c., can only be obtained after considerable practice, an attempt has been made in every case, by the employment of solutions of *definite* strength for *definite* times, to enable even a beginner to arrive at an average result without loss of time in experimenting.

It is therefore hoped that the usefulness of the work as a compact and inexpensive book of reference for the experienced worker, as well as a guide to those commencing investigations, may justify its publication.

Very few "double-stains" have been given, as even when good results could be obtained by using two dyes in the same solution, better and more precise staining was effected when the same dyes were used separately one after the other.

Amongst the stains but little known, which have given very good results, may be mentioned Benzopurpurine as a counter-stain to Hæmatoxylin; Congo Red for the central nervous system; and Rose Bengale for the detection of early Amyloid. The last reaction has not been previously noticed.

For the pathological work, the co-operation of F. J. Warwick, M.B. (Cantab), a practical microscopist of much experience in this subject, has been most valuable.

Most of the methods given under Micro-organisms have been translated from the original papers. Kühne's methods are taken from the translation by Dr. Vincent Harris.

P. W. S.

413, OXFORD STREET.
April 5, 1892.

CONTENTS.

	PAGE
Hardening and Fixing Reagents	1—10
Decalcifying Fluids	11—13
Embedding Media	14—19
Nuclear Stains	20—39
Plasmatic Stains	39—42
Specific Stains	43—48
Specific Methods	48—51
Staining Cellulose and its Modifications	52—56
Staining Sieve-areas	56
Micro-organisms	57—77
Stains—Intermediate Reagents—General Methods—Special Methods.	
Dehydration	78
Clearing Agents	79—82
Mounting Media	83—86
Normal or Indifferent Fluids	87
Dissociating Fluids	88—90

HARDENING AND FIXING REAGENTS.

For general histological and pathological work, the hardening agents usually employed are Alcohol and the Bichromate solutions. The others are only used for special applications.

Alcohol will not penetrate well pieces larger than a cubic inch, while the Bichromates are applicable to material of any size.

If rapid hardening is required, Alcohol, Erlicki's Fluid, or Klein's Fluid is employed, but the two latter have a tendency to make the tissues brittle if they remain in the solution too long.

A relatively large quantity of hardening fluid, about 20 times the bulk of the object to be hardened, should always be used, and the pieces should be as small as circumstances will allow. It is best to suspend the object near to the top of a tall vessel filled with the fluid, which is accomplished by placing the pieces of tissue on absorbent wool enclosed in a muslin bag, but this is not always considered to be necessary. The pieces can simply be placed in a bottle with the fluid, which should be circulated from time to time, and changed when necessary.

As soon as the objects are sufficiently hard they should be removed from the hardening solution and placed in 70 p. c. Alcohol till required.

Of the special fixing agents Osmic Acid is the best, more

particularly in combination with Chromic Acid and Acetic Acid, as in Flemming's and Fol's solutions.

Fresh vegetable tissues (leaves, roots, and stems) are best fixed in Alcohol. In the majority of instances 90 p. c. Alcohol will answer the purpose; but where the nuclei in particular are to be examined, a stronger Alcohol (Absolute) should be used.

Fresh-water algæ, &c., can be fixed in Picric Acid or a Chrom-Osmic solution (Flemming's or Fol's); but, if it is desired to preserve the colour of the chlorophyll, the Acetate of Copper solution is recommended.

Sections of fresh tissues are sometimes fixed and stained at the same time by Acetic Acid mixed with Methyl Green or Gentian Violet.

Most stems, barks, and roots can be cut after fixing in 90 p. c. Alcohol; but when very hard they can be softened by removal from the Alcohol into Thymol Water, or equal parts of Thymol Water and Glycerine. In rare cases it is necessary to use 2 to 5 p. c. of Caustic Potash in the Thymol Water.

(i.) **ALCOHOL**.

Suitable for all tissues except the nervous system and tissues undergoing fatty infiltration or degeneration; from these the fat would be dissolved out to a considerable extent by the Alcohol.

Glandular structures and tissues to be examined for Bacteria are placed at once into 90 per cent. Alcohol, or Absolute Alcohol, but in other cases, where it is required to preserve the normal structure of the tissues, they are at first placed in 50 per cent. Alcohol for a day or two, depending upon the size of the pieces, then into 70 per cent. Alcohol for a similar time, and subsequently into 90 per cent. Alcohol until the hardening is complete.

For fixing nuclear figures the strongest Alcohol (Absolute) should be used, and the pieces should be as small as possible.

Throughout this book, per cent. of Alcohol is given by weight; when Methylated Spirit is used the gravities may be slightly different. The following formulæ are sufficiently near for all practical purposes.

Absolute Alcohol, about 98 per cent. Sp. gr. 0·797.
The commercial article. Alcohol Ethylicum B.P.

Alcohol, 90 per cent. Sp. gr. 0·823.
Absolute Alcohol, 14 vols.; Distilled Water, 1 vol.: mix.

The strongest commercial **Methylated Spirit** (64 o. p.) is practically the same strength as 90 p. c. Alcohol, and being duty free is much cheaper and answers every purpose.

This, of course, only applies to Methylated Spirit free from Mineral Naphtha. The ordinary "Retailer's" quality contains Naphtha, and becomes turbid when mixed with water.

Alcohol, 84 per cent. Sp. gr. 0·838.
This is Rectified Spirit B.P. (56 o. p.).

Alcohol, 70 per cent. Sp. gr. 0·872.
Absolute Alcohol, 3 vols.; Distilled Water, 1 vol.
Rectified Spirit, 6 vols.; Distilled Water, 1 vol.
Methylated Spirit, 4 vols.; Distilled Water, 1 vol.

Alcohol, 50 per cent. Sp. gr. 0·918.
Absolute Alcohol, 5 vols.; Distilled Water, 4 vols.
Rectified Spirit, 5 vols.; Distilled Water, 3 vols.
Methylated Spirit, 5 vols.; Distilled Water, $3\frac{1}{2}$ vols.

(ii.) **ALCOHOL AND ACETIC ACID.**
Absolute Alcohol, 75 c.c.; Glacial Acetic Acid, 25 c.c.
This is an excellent fixing agent for nuclei. Tissues are

placed in it from 6 to 12 hours, and then transferred to 90 p. c. Alcohol until hardened. When the hardening is completed, transfer the tissue to 70 p. c. Alcohol till wanted.

(iii.) **CHROMATE OF AMMONIUM.**

5 p. c. aqueous solution.

Especially useful for delicate cell structures, as parenchyma of secreting cells. Tissues are placed for 48 hours in this reagent, thoroughly washed in Distilled Water and passed through diluted to strong Alcohol.

(iv.) **BICHROMATE OF AMMONIUM.**
BICHROMATE OF POTASSIUM.

2 to 5 per cent. aqueous solution of either.

Simple Bichromate solutions are sometimes preferred to Müller's Fluid for hardening parts of the central nervous system, but opinions vary as to their greater usefulness even for this purpose. A two per cent. solution should be used to start with and changed after 24 hours and again after 3 days, and then once a week for the next 6 weeks, increasing the strength by one per cent. each week till it reaches five per cent. In this way an entire brain may be hardened thoroughly and satisfactorily in 8 weeks.

After hardening for a month in Müller's Fluid, it is often advantageous to finish the process by a fortnight's immersion in five per cent. Bichromate solution.

(iv.*a*) **MULLER'S FLUID.**

Bichromate of Potassium, 2 grms.; Sulphate of Sodium, 1 grm.; Water, 100 cc.

A good hardening reagent for delicate structures where the process is required to be gradual. It is well adapted for the eye. It penetrates evenly and thoroughly, so that large pieces can be hardened in it, while at the same time it has no tendency to over-harden and render the tissues brittle.

The disadvantages attaching to all the Bichromate solutions are that they discolour the tissue more or less. A long time is required for the hardening, and if tissues be kept in them too long they sometimes produce netlike forms of coagulation, or deposits of a dark granular precipitate in the cells and intercellular substances; they are apt to develop moulds. They dissolve out calcareous deposits.

When the nervous system is stained by Weigert's method, the tissue must be hardened in this or one of the other Bichromate solutions.

It is unsuitable for hardening tissues to be examined for the amyloid reactions or for bacteria.

Tissues ought to be placed in at least 20 times their bulk of "Müller." This must be changed after 24 hours, again after 3 days, and then once a week for the next 6 weeks. Even for small pieces 6 weeks in all are generally required. The tissues, after removal from "Müller," are soaked in 50 p. c. Alcohol from 1 to 2 days, then 70 p. c. Alcohol for a similar time, and finally placed in 90 p. c. Alcohol, which should be changed occasionally so long as it becomes coloured. In the case of large masses, a preliminary washing in water will save Spirit and do no harm.

(iv.*b*) **MULLER'S FLUID AND SPIRIT.**

Müller's Fluid, 3; 90 p. c. Alcohol, 1.

This is usually directed to be cooled before using it, but as there is only a rise of 11° F. on mixing them, the precaution is hardly necessary.

Used in the same cases as "Müller," only its action is more rapid. It takes about half the time of "Müller" to harden tissues. Whilst hardening in this, and during the subsequent washing in Alcohol, tissues should be kept in the dark.

The other Bichromate solutions may be used with Spirit in the same way.

(iv.*c*) **ERLICKI'S FLUID.**

Bichromate of Potassium, 5 grms.; Sulphate of Copper, 1 grm.; Water, 200 cc. This solution hardens quicker than "Müller," and is used in much the same way.

It is sometimes stated that this solution *should always be prepared fresh when required*, but this is not necessary as it undergoes no change.

This is the best fluid for hardening large organs quickly. It will harden a spinal cord in 10 days, and is the most suitable fluid for human embryoes.

It tends to render the tissues brittle, and also gives the sections a greenish tinge.

(v.) **CHROMIC ACID.**

10 p. c. aqueous solution is kept for stock; for use it is diluted with 20, 30, or 50 times its bulk of Water, according to the consistence and size of the tissue. The larger or firmer the tissue the weaker the solution.

The pieces to be hardened must not be larger than $\frac{1}{2}$ cubic inch; the fluid must be changed every day for the first 3 days, and then every third day. Tissues are hardened in from 1 to 14 days, according to their nature and size; they are then well washed in Water and kept in 70 p. c. Alcohol in a dark place to avoid deposits.

Chromic Acid is energetic, but it does not penetrate well, and it makes the tissues brittle. Tissues thus hardened are not well stained by Carmine; Hæmatoxylin or Safranine should be used.

(v.a) **KLEIN'S FLUID.**

Chromic Acid solution (10 p. c.) 1 cc.; mix with Water 60 cc.; then add 90 p. c. Alcohol, 30 cc.

The Alcohol is added to increase the penetration of the Chromic Acid. This solution should be made fresh as required and kept from the light, as the Chromic Acid is rapidly reduced by the Spirit.

Suitable for all tissues, but more particularly the nervous system. The fluid must be changed every day for the first three days, and then every third day. The hardening is complete in about a week to a fortnight.

In all cases except the nervous system, the tissue after hardening must be well washed in Water, and preserved in 70 p. c. Alcohol.

The nervous system should be transferred direct from Klein's Fluid to the Alcohol.

(v.b) **RABL'S FLUID.**

Chromic Acid solution (10 per cent.), 7 cc.; Water, 200 cc.; Formic Acid (sp. gr. 1·2), 5 drops.

Fresh tissues in small pieces are placed in it from 12 to 24 hours; they are then washed in Water and hardened in Alcohol of gradually increasing strength.

This solution is especially useful for demonstrating cell division (mitosis) and nuclei.

The following are used more particularly as **fixing** agents.

(vi.) **CORROSIVE SUBLIMATE.**

5 p. c. aqueous solution.

This is a capital fixing agent, and acts quickly.

Tissues ought to be placed in the solution from 15 minutes to 2 hours, and then in. 90 p. c. Alcohol to complete the hardening.

After this the tissues must be washed in 70 p. c. Alcohol; if Water be used black specks are apt to form.

The action of the fluid can be accelerated by heating it to 38° C. (100·4° F.)

For glands and glandular structures the following is very useful:—Corrosive Sublimate, 10 grms.; Alcohol (70 p. c.), 100 cc.; Glacial Acetic Acid, 1 cc. Small pieces are placed in this for an hour and the hardening completed in Alcohol.

(vii.) **NITRIC ACID (Altmann).**

3 p. c. aqueous solution. Sp. gr. about 1·02. Used as a fixing agent.

The pieces of tissue must be as small as possible, and placed in the Acid until quite fixed; this takes from 15 to 30 minutes. Useful for fixing nuclei of cells, embryoes and the retina.

viii.) **OSMIC ACID.**

1 p. c. aqueous solution, which must be carefully preserved from dust, as it is readily reduced (blackened) by small quantities of organic matter. This solution is usually kept in the dark.

Especially useful for delicate structures, when rapid fixing is required.

For use it is diluted with 4 or 5 times its volume of Distilled Water, or more generally in conjunction with Chromic Acid as in Flemming's and Fol's solutions. Tissues are placed in it from 8 to 10 hours, well washed in Distilled Water, cut, and mounted in Farrant's solution.

Some authorities prefer to fix with the *vapour*; this is done by suspending the tissue in a bottle containing some solution or crystals of Osmic Acid.

Tissues containing Fat are blackened by Osmic Acid, *see Specific Stains.*

(viii.*a*) **FLEMMING'S FIXING SOLUTION.**

Osmic acid (1 p. c. solution), 80 cc.; Chromic Acid (10 p. c. solution), 15 cc.; Glacial Acetic Acid, 10 cc.; Distilled Water, 95 cc.

This and the following solution are used solely as fixing agents. Tissues are placed in either for about twelve hours; they are then thoroughly washed in Water and hardened in Alcohol of gradually increasing strength (*see Alcohol*).

Especially useful for fixing tissues undergoing cell division (mitosis). This fluid need not be used in the dark.

Sections should be stained with Safranine, Hæmatoxylin, or Gentian Violet.

(viii.*b*) **FOL'S SOLUTION.**

Osmic Acid (1 p. c. solution), 4 cc.; Chromic Acid (10 p. c. solution), 5 cc.; Glacial Acetic Acid, 10 cc.; Distilled Water, 181 cc.

Used for the same purpose as the preceding.

Flemming's and Fol's solutions represent the maximum and minimum strength of Chrom-Osmic mixtures for fixing purposes; various intermediate strengths are also employed.

(ix.) **PICRIC ACID.**

A saturated aqueous solution (1 in 75), or 1 p. c. solution in 70 p. c. Alcohol, are used for fixing and hardening. The action is rapid, taking from 1 to 24 hours.

After hardening, the tissue is washed in 70 p. c. Alcohol

(not Water) and preserved in 90 p. c. Alcohol. Picric Acid solutions dissolve out calcareous deposits.

(x.) PICRO-SULPHURIC ACID (Kleinenberg).

Saturated aqueous solution of Picric Acid, 100 cc.; pure Sulphuric Acid, 2 cc.; filter; to each 100 cc. of filtrate add 300 cc. of Distilled Water.

<small>As in Kleinenberg's formula four-fifths of the Picric Acid is thrown out of solution, the same final result can be obtained by the following:—</small>

Saturated Solution of Picric Acid, 20 cc.; Distilled Water, 380 cc.; Sulphuric Acid, 2 cc.

Especially useful for the rapid fixing of very soft tissues, as sarcoma and myxoma; also for preserving embryoes. It is used for fixing and preserving the earth-worm.

Tissues are placed in the fluid from 3 to 12 hours; they are then washed in 70 p. c. Alcohol, and finally placed in 90 p. c. Alcohol.

Carmine is the best stain to use after this solution; but if Hæmatoxylin be used, the sections must first be placed in a solution of Carbonate of Lithium, and then washed in Water.

(xi.) ACETATE OF COPPER.

Acetate of Copper, 1 grm.; Distilled Water, 200 cc.; Glacial Acetic Acid, 1 cc.; Glycerine, 200 cc.; Corrosive Sublimate, 3 grms.

This solution is especially useful for fixing and mounting green Algæ.

DECALCIFYING FLUIDS.

In the case of bony structures or tissues so impregnated with Lime Salts as to interfere with section cutting, the material must be decalcified in one of the following solutions. The general plan is to combine an Acid capable of dissolving out the mineral matter with a hardening agent which will prevent swelling of the tissues.

Hydrochloric Acid is generally used with Alcohol. Diluted Nitric Acid may be used alone, but usually in conjunction with Alcohol, Chromic Acid, or an equivalent quantity of Potassium Bichromate.

The older the bone the stronger will be the acid required.

It is advisable to harden first in Alcohol or Bichromate before decalcifying, except in the case of Ebner's solution where the two processes go on together.

(i.) **ARSENIC ACID.**

4 p. c. aqueous solution, used at a temperature of 30° to 40° C.

It acts rapidly. Tissues after softening should be kept in Alcohol.

(ii.) **CHROMIC ACID.**

0·1 to 0·5 p. c. aqueous solution.

Used for softening small pieces of bone, but when use alone it causes much shrinkage in the tissues.

It is best used in conjunction with Nitric Acid.

(iii.) **HYDROCHLORIC ACID.**

1 to 10 p. c. in water of the strong acid (sp. gr. 1·16.) Softens rapidly, but causes tissues to swell and become almost structureless. This is remedied by the addition of Alcohol as in the following :—

(iv.) **EBNER'S SOLUTION.**

90 p. c. Alcohol, 1000 cc.; Water, 200 cc.; Chloride of Sodium, 5 grms.; Hydrochloric Acid (sp. gr. 1·16), 5 cc.

In this solution the Spirit and Salt are added to prevent the swelling of the tissue. This solution takes a long time to soften, its action being very gradual, requiring about six weeks. To each volume of the tissue 200 to 300 vols. of the solution must be used; when softening large bones it is necessary to add a small quantity of Hydrochloric Acid from day to day, to replace that which combines with the Lime Salts, and examine the tissue with a needle from time to time, till the process of softening is complete; the tissue is then washed in a large volume of Water to remove the acid, and finally kept in 70 p. c. Alcohol.

(v.) **HYDROCHLORIC ACID AND GLYCERINE.**

Glycerine, 95; Hydrochloric Acid, 5.

This is especially useful for softening teeth; it softens and yet does not in any way interfere with the structure or bony appearance; it also acts as a preservative. Its action is slow, taking several weeks.

(vi.) **NITRIC ACID.**

Aqueous or Alcoholic solution containing from 1 p. c. for small bones, up to 5 p. c. for large heavy bones.

They are first placed in 90 p. c. Alcohol for two or three

days, then treated with the Nitric Acid solution, which is changed daily for about a week; as soon as decalcification is complete, wash thoroughly in Water, and place in 70 p. c. Alcohol.

(vii.) **NITRIC ACID AND CHROMIC ACID.**

Chromic Acid, 1 grm.; Water, 200 cc.; Nitric Acid 3 cc.

This is one of the best softening fluids for general use, the Chromic Acid hardening the bone protoplasm, while the Nitric Acid dissolves out the calcareous salts.

It takes about a fortnight, and the solution must be renewed every third day. The tissue is then well washed in water and transferred first to weak and then to strong Alcohol.

(viii.) **PICRIC ACID.**

A saturated aqueous solution. Useful for softening fœtal bone. The tissue should be suspended in twenty times its bulk of fluid, which must be kept saturated with crystals of Picric Acid. It takes about a month; the tissue when softened must be first well washed in Water and then passed from weak into strong Alcohol (*see* "Alcohol").

EMBEDDING MEDIA.

These are mostly used to support the tissues during the cutting with a microtome.

The simplest method is that adopted for firm vegetable tissues. A cylinder of carrot is made to fit the microtome, and then divided vertically. The piece of tissue is fitted in between the two halves of carrot; the carrot and tissue are cut at the same time, the former, of course, transversely. The direction of the latter will depend upon the position in which it is placed in the carrot. Another method is to embed in Paraffin, and fit the block of Paraffin to the microtome.

Neither of the above are suited for delicate or porous tissues. They must be supported in the interstices by saturation with the embedding material. For this purpose Gum, Celloidin, and Paraffin are used. When it is desired to support in position tissues which would otherwise fall to pieces when cut into sections, Celloidin is generally used. Saturation with Paraffin is mostly employed for serial cutting.

(i.) **GUM (Mucilage).**

Colourless Gum Acacia, 2; Cold Water, 3.

Dissolve the Gum in the Water, and to each ounce of the solution add 10 grains of Carbolic Acid.

(ii.) **GUM AND SYRUP.**

Mucilage of Acacia, 50 cc.; Simple Syrup B.P., 30 cc.

EMBEDDING MEDIA.

Method of Using.

Before using either of the above for embedding, the tissues must be washed for several hours in frequent changes of Water, to free them from Alcohol, and then placed for 24 hours (preferably 3 or 4 days) in the Gum or Gum and Syrup, to saturate them before freezing on the microtome.

(iii.) CELLOIDIN.

Celloidin dissolved in equal parts of Absolute Alcohol and Ether, an 8 p. c. Solution being used for **Thick Celloidin**, and diluted with an equal volume of the Ether-Alcohol for **Thin Celloidin**.

Method of Using.

The tissues (previously hardened by one of the methods already given), if not already in strong Alcohol, should be placed there; they are then immersed for 12 hours in equal parts of Absolute Alcohol and Ether. After this they are immersed in Thin Celloidin for 24 hours; then for another 24 hours (preferably 2 or 3 days) in Thick Celloidin, so that it may thoroughly permeate the tissues. The pieces are now removed from the solution, and placed on small cubes of soft wood, care being taken to see that the wood is quite dry. Allow the Celloidin to harden for a few minutes, and then place the wood with the tissue in 90 p. c. Alcohol for 12 to 24 hours. They are then ready for clamping in the microtome.

If material, saturated with Celloidin, is to be cut on the freezing microtome, the piece when removed from the Thick Celloidin, is placed with some of the Celloidin in a small paper case. Allow a thin film to form on the surface by evaporation, and then place the paper case with contents into 80 or 90 p. c. Alcohol for 12 to 24 hours or longer to harden. When it assumes the consistency of soft cheese take it out of the Alcohol, remove the paper case, and place it in Water to wash away superfluous Alcohol; the piece is then dipped in Mucilage, and finally frozen on the microtome in the usual way.

(iv.) GLYCERINE GELATINE (Kleb's).

Best Gelatine (French), 10 parts; wash and allow this to stand in Distilled Water till it swells up: pour off excess of Water, melt at a gentle heat, and add 10 parts of Glycerine, and 2 p. c. of Carbolic Acid to preserve it.

Method of Using.

The tissue, which has previously been well washed with Water to remove Alcohol, is saturated with the melted Glycerine Jelly, which is then allowed to set. This may now be frozen between pieces of carrot on the microtome by means of Gum; or it may be clamped between carrot for cutting.

This method is useful for flat structures, such as the walls of hollow organs, or membranes.

(v.) PARAFFIN.

Paraffin melting at between 45° C. (113° F.) and 50° C. (122° F.) according to the temperature of the room, and the nature of the material to be cut.

This medium is best adapted for embedding tissues that have been stained in bulk.

Method of Using.

Some tissues are *embedded* in Paraffin, or other suitable substance, only as a means of holding them in the microtome; other tissues which are very porous or brittle require to be *saturated* with the embedding material to hold the parts together during or after the cutting.

Embedding.—The tissue which has been dehydrated is dipped for a second or two into melted Paraffin, and taken out again. This gives the piece a coating of Paraffin. The tissue may also be transferred direct from the Alcohol to the cell of Paraffin. In either case the piece is put into the melted Paraffin, which has been poured into a cell, and which has cooled sufficiently to hold the piece in any position required. Another method is to make a hole in a small block of Paraffin, fit in the piece of tissue, and fill up with warmed Paraffin.

The cell to hold the Paraffin can be made in a variety of ways. A small rectangular tray can be made by folding paper or thin card; a cylinder can be formed by rolling paper round a cork (allowing the paper to project beyond the cork) and fixing it with a pin run through the paper into the cork. Cells can also be formed by means of two pieces of brass, each of them being bent at right angles, and the cylinders can be cut from brass-tubing.

When the sections have been cut, the Paraffin can be removed by immersion in Naphtha, Benzol, Toluol, or Xylol, all of which are good solvents of Paraffin.

For staining, the sections must then be passed through Absolute Alcohol to remove the liquid hydrocarbon.

Saturation.—The piece of tissue, after staining in bulk, is placed in strong Spirit and afterwards in Absolute Alcohol. It is then transferred to Cedar Oil or Xylol, and when cleared is placed in a suitable (not too large) quantity of Paraffin, kept just above the melting point, until the tissue is saturated with it. Both Cedar Oil and Xylol are convenient media, as either mixes readily both with Absolute Alcohol and melted Paraffin. If the pieces of tissue carry much oil into the Paraffin, a second bath of the latter may be used. When the saturation is complete (2—5 hours), the Paraffin should be quickly cooled, and a piece containing the tissue can be cut out with a warmed knife. It is then ready for mounting on the microtome.

Another Method of Saturation.—The tissue after staining is passed into 90 p. c. Alcohol, and then into Absolute Alcohol, in which it must remain for not less than 3 hours. It is next placed in ten times its bulk of Chloroform or Benzol until thoroughly saturated; then small pieces of the Paraffin are added to the Chloroform or Benzol until no more will dissolve; in this the tissue is allowed to remain for 2 or 3 hours. Gentle heat is then applied to drive off the solvent and melt the Paraffin, without injuring the specimen. The mass is allowed to set, and gentle heat is again applied to the vessel containing the mass to loosen it. The superfluous Paraffin is trimmed and the mass fused on to the socket of the rocking microtome.

(vi.) **RESIN AND WAX.**

Resin, 3; Yellow Wax, 1; melt together.

It is used for holding tissues (bone and teeth) whilst being ground.

Pieces are cut as thin as possible with a saw, and one side is ground on an oilstone; the piece is then placed in the Resin and Wax (liquefied by heat) and cemented on the ground side to a piece of glass. When hard the tissue is rubbed down to the required thickness on an oilstone. The Resin and Wax can be removed from the tissue by Oil of Turpentine, in which both are soluble.

PRESERVING SECTIONS.

Sections are best preserved (previous to staining and mounting) in 50 p. c. Alcohol, or in a mixture of Glycerine and Thymol Water (1 in 1500) in equal volumes. If kept long in strong Alcohol they are apt to shrivel.

MANIPULATION OF SECTIONS.

Thin sections which have been kept for some time in Alcohol are apt to get folded or crumpled. If these be placed one at a time on the surface of some water, they will open up immediately, with the assistance if necessary of a pointed glass rod (glass needle.) Glass rods (about $\frac{1}{4}$-inch diameter for a $\frac{1}{2}$-inch section) are also convenient to transfer sections from one dish or bottle to another in the staining process, the section being allowed to wrap itself *round* the rod. When a section is being treated in strong Alcohol or any liquid which tends to make it shrink or curl, it should be allowed to remain wrapped round the rod, which prevents any change of shape. In transferring a section from the rod to a slide, it is only necessary to place on the slide a drop or two of the liquid from which it is being removed. Lay the rod with the adhering section on the wetted slide, and *roll* the section off the rod flat on to the slide. Very delicate sections can be manipulated in this way after a little practice.

Vegetable sections are usually thicker than Animal sections, and can be manipulated with a lifter, or a pair of bent forceps the ends of which are broad and flattened.

Washing sections in Alcohol, dehydrating, and clearing, are conveniently performed in small wide-mouthed bottles.

STAINS.

These may be divided into **Nuclear, Plasmatic,** or **Specific,** according to their power of staining the nuclei, the whole tissue, or certain elements of the tissue only.

Nuclear Stains.—Hæmatoxylin, Carmine and Cochineal, Methylene Blue, Methyl Green, Safranine, Vesuvine (Bismarck Brown), Gentian Violet, Hoffmann's Blue (acidified), and Fuchsine.

Plasmatic Stains.—Picric Acid, Benzopurpurine, Eosine, Erythrosine, Orange, and Acid Fuchsine.

Specific Stains.—Osmic Acid, Chloride of Gold, Nitrate of Silver, Nigrosine, Acid Fuchsine, Congo Red, Dahlia Violet, Methyl Violet, Iodine, Rose Bengale, Safranine, and Victoria Blue.

Stains and methods for **Micro-Organisms** are treated separately.

NUCLEAR STAINING.

There is no colouring matter which is a **nuclear** stain only. All of them have more or less effect upon the ground substance also. The value, therefore, of any dye as a nuclear stain depends upon the **comparative** affinity which it has for nuclei and ground substance respectively.

Direct Nuclear Stains are those in which the nuclei are dyed so rapidly, compared with the rest of the tissue, that no particular treatment is necessary to differentiate them.

Indirect Nuclear Stains are those in which the whole tissue is deeply stained, and the nuclei are differentiated by subsequent removal of the colour from the ground substance.

For histological and pathological work, Hæmatoxylin is beyond doubt the best stain for general use. The nuclei are clearly defined, the manipulation is easy, and good results may be obtained with less experience than by any other method.

With Ehrlich's solution it is easy to stain the nuclei without affecting the ground substance to any marked degree; whereas with Carmine and the Aniline dyes the two are stained together, and it is less easy to remove effectually the colour from the ground substance and leave the nuclei sharply defined.

When the tissues have been hardened in the Chromates, the advantage of Hæmatoxylin over Carmine in rapidity of staining is very marked. Carmine, on the other hand, possesses advantages in staining the nervous system, to which Hæmatoxylin is inapplicable, except by the somewhat tedious process of Weigert or one of its modifications.

HÆMATOXYLIN.

This is essentially a nuclear stain, as it selects the nuclei first; but if the solution be too strong, or the action be continued too long, it will also stain the ground substance. It is very extensively employed, and always in conjunction with Alum, except in the special uses to which it is put in Weigert's method for the nervous system, in Heidenhain's method for staining in mass, and Benda's copper method.

Of the solutions, Ehrlich's and Delafield's are recommended for sections, and Kleinenberg's for staining in mass.

Sections are counter-stained with Benzopurpurine, Eosine, Erythrosine, or Rubin and Orange.

Hæmatoxylin has been objected to on the ground that the solutions deteriorate by keeping, and that the stained sections are not permanent. These objections certainly do not apply to Ehrlich's solution, a quantity of which, made five years ago, is now in perfect order, and sections are still good which were stained and mounted four and five years since.

In order to change the violet colour of the sections into blue it is usual to soak them for some hours in "tap water," on the supposition that the alkalinity of the Carbonate of Lime, dissolved in most natural waters, was necessary to the change. That this is not the case is evident from the fact that sections stained in a well-matured Ehrlich's Hæmatoxylin solution (which of course is itself acid) will become blue in pure Distilled Water, freed from even the slightest trace of Ammonia. The change, however, is undoubtedly assisted by a trace of Alkali, and when sufficient Lime Carbonate does not exist in the water, uniformly good results can always be obtained in a few minutes by using **Distilled Water containing 10 grains of Bicarbonate of Sodium to the pint (20 ounces)**. Light is in no way essential to the development of the blue colour, and direct sunlight should be carefully avoided on account of its bleaching action.

After the blue colour is developed, transfer the sections to 70 p. c. Alcohol, as the colour fades much more quickly when kept in water.

It is occasionally stated that **previous to staining** the sections must be freed from every trace of acid, as otherwise the colour will not be permanent. Considering that one of the best stains is itself acid, and that Acidulated Water or Alcohol may freely be used to remedy overstaining, it is much more to the point that sections, **after** the staining process is completed, should be carefully freed and preserved from any trace of acidity, more particularly when using counter-stains. Freshly prepared solutions of Hæmatoxylin stain badly and diffusely. When exposed to the air in strong daylight they undergo a change technically called "ripening," after which the staining is much more precise.

The ripening process appears to depend upon conversion of part of the Hæmatoxylin into Hæmatein by absorption of oxygen from the air in presence of traces of free Ammonia. By dissolving Hæmatoxylin crystals in Proof Spirit with $\frac{1}{5}$ of their weight of Ammonium Carbonate, and exposing to the air in a shallow dish, the "ripening" process will be more complete in 24 hours than after 3 months' exposure in the usual way.

As solutions, even when made from the same formula, differ widely in their staining power according to the ripening which the Hæmatoxylin has undergone, no very definite directions as to dilution and time of immersion can be given. As a rough guide, it may be stated that a well-ripened Ehrlich's solution, diluted 1 to 4 of Distilled Water, and Delafield's diluted 1 to 10, will stain well in about 20 minutes.

Accidental overstaining may be remedied by immersing the sections in 70 p. c. Alcohol, to which $\frac{1}{10}$ or $\frac{1}{5}$ per cent. of strong Hydrochloric Acid has been added. When the overstaining is very deep, $\frac{1}{2}$ p. c. Acid may be used to save time, but requires caution. The sections are subsequently washed in a dilute aqueous solution of Bicarbonate of Sodium (1 grain in 2 ounces).

(i.) **BÖHMER'S HÆMATOXYLIN.**

(*a*) Hæmatoxylin, 1 grm.; Absolute Alcohol, 10 cc.

(*b*) Alum (Ammonia), 10 grms.; Distilled Water, 200 cc.

Mix the two solutions, and after a week filter.

This is probably the earliest formula, as Hæmatoxylin staining was introduced by Böhmer.

NUCLEAR STAINS.

(ii.) EHRLICH'S HÆMATOXYLIN.

Hæmatoxylin, 2 grms.; Absolute Alcohol, 100 cc. Dissolve and add—Glycerine, 100 cc.; Distilled Water, 100 cc.; Alum (Ammonia), 2 grms.; Glacial Acetic Acid, 10 cc.

It should be exposed to daylight for at least a month, before use, removing the stopper at intervals.

(iia.) EHRLICH'S HÆMATOXYLIN (Ammoniated).

Hæmatoxylin, 2 grms.; Ammonium Carbonate, 0·4 grms.; Proof Spirit, 40 cc.: dissolve the Ammonium Carbonate and the Hæmatoxylin in the Proof Spirit and expose to the air in a shallow dish for 24 hours, make up the volume to 40 cc. with Proof Spirit (warming if necessary to dissolve any separated crystals) and add Ammonia Alum, 2 grms., dissolved in Distilled Water, 80 cc.; Glycerine, 100 cc.; Rectified Spirit, 80 cc.; Glacial Acetic Acid, 10 cc. This solution requires no "ripening" and is ready to be diluted for use (1—10).

(iii.) DELAFIELD'S HÆMATOXYLIN.

To 400 cc. of a saturated aqueous solution of Ammonia Alum add 4 grms. of Hæmatoxylin dissolved in 25 cc. of Absolute Alcohol; leave the solution exposed to the light and air in an unstoppered bottle for 3 or 4 days; filter, and add to the filtrate 100 cc. of Glycerine and 100 cc. of Methylic Alcohol (Wood Spirit); allow the solution to stand in the light until it is a dark colour, re-filter and preserve in a stoppered bottle.

Ammonia Alum dissolves about 1 in 11 of Water.

(iv.) HAMILTON'S HÆMATOXYLIN.

Hæmatoxylin, 12 grms.; Alum (Ammonia), 50 grms.; Glycerine, 65 cc.; Distilled Water, 130 cc.; boil, and while hot add 5 cc. of liquid Carbolic Acid. Allow the mixture to stand in daylight for at least a month to ripen before using.

It is five to ten times stronger in Hæmatoxylin than the other solutions, and the greater part of the Alum (85 p.c.) crystallizes out on cooling.

When this solution is employed sections must be immersed for a shorter time, or it must be more diluted than the other solutions.

(v.) KLEINENBERG'S HÆMATOXYLIN.

The peculiarity of this solution is that it contains a quantity of Chloride of Calcium, which, when staining in bulk, sets up diffusion currents between the alcohol in the material to be stained and the alcoholic staining solution, so enabling the latter to penetrate more rapidly.

Great differences exist in the formulæ given for this solution. It would appear from the *Quarterly Journal of Microscopic Science*, 1879, p. 208, that Kleinenberg himself latterly used the following process:—

"Prepare a saturated solution of Calcium Chloride in 70 p. c. Alcohol, with the addition of a little Alum; after having filtered, mix one volume of this with 6 to 8 volumes of 70 p. c. Alcohol. At the time of using the liquid pour into it as many drops of a concentrated solution of Hæmatoxylin in Absolute Alcohol as are sufficient to give the required colour to the preparation, of greater or less intensity according to desire."

In an older formula, given by Balfour and Foster in their "Elements of Embryology," and copied into various text-books, a saturated solution of Alum in 70 p. c. Alcohol is used instead of 70 p. c. Alcohol alone, for diluting the strong Chloride of Calcium solution; but as 70 p. c. Alcohol only dissolves 1 in 1000 of Alum, this serves no useful purpose. It is also directed that the Calcium Chloride solution should be *saturated* with Alum; but this is quite impracticable. Double decomposition takes place, to such an extent that when saturated with Alum the liquid actually *solidifies* from separation of Sulphate of Lime.

As Kleinenberg's above formula is still very indefinite in regard to the quantity of Alum used, and as the extent of the decomposition will vary greatly with the fineness of the Alum powder, the following modification is recommended as giving an exact and uniform result.

Hæmatoxylin, 2½ grms.; Crystallized Calcium Chloride 20 grms. in 10 cc. of Distilled Water; Alum, 3 grms. in 16 cc. of Distilled Water; Rectified Spirit, 240 cc.

Dissolve the Calcium Chloride and the Alum in their respective quantities of water, by the aid of heat; mix the solutions and immediately dilute with the Rectified Spirit; after an hour filter and add the Hæmatoxylin.

The above quantities make a good working solution. Any dilution which is required must be made with some more of the filtered liquid to which Hæmatoxylin has not been added.

Crystallized Chloride of Calcium is preferred to the ordinary anhydrous variety, on account of its greater purity: the anhydrous Chloride is always alkaline. Whichever be used, a saturated solution in 70 p. c. Alcohol will contain about 40 p. c. of $CaCl_2$.

As in the finished solution, the whole of the Aluminium exists as Chloride, it has been proposed to avoid the double decomposition with Alum, by simply adding to the Chloride of Calcium solution a definite quantity of Aluminium Chloride. It is difficult, however, to obtain this latter free from uncombined Hydrochloric Acid, so that the double decomposition process is the best.

Kleinenberg adds the Hæmatoxylin just before use. Lee says it is best to add it 24 to 48 hours previous to use, with the additional note that the solution does not keep; while Stirling says it should be kept for at least a month before it is wanted.

The solution is principally used for staining in bulk, and as a rule the larger the object the weaker in Hæmatoxylin should the solution be made to stain slowly and uniformly.

(vi.) **WEIGERT'S HÆMATOXYLIN** for the central nervous system. (*See p.* 48.)

Hæmatoxylin, 1; Absolute Alcohol, 10; Distilled Water, 90; aqueous solution of Lithium Carbonate (1 in 70), 1.

(vii.) **RENAUT'S GLYCERINE HÆMATOXYLIN.**

Saturate Glycerine with Potash Alum, and to it add drop by drop a saturated solution of Hæmatoxylin in 90 p. c. Alcohol to form a deeply coloured solution; let it stand exposed to daylight for at least a week, and filter.

Both this and the following solution are employed for

mounting unstained sections. After some time the sections absorb the colour and become stained.

(viii.) **RENAUT'S EOSINATED GLYCERINE HÆ-MATOXYLIN.**

Concentrated aqueous solution of Eosine, 30 cc.; saturated solution of Hæmatoxylin in Alcohol, 40 cc.; saturated solution of Potash Alum in Glycerine, 130 cc.; mix and let it stand in a beaker for 5 or 6 weeks, so that the Alcohol may evaporate, but protected from dust, then filter.

If desired, the filtrate can be diluted with Glycerine.

(ix.) **HEIDENHAIN'S HÆMATOXYLIN METHOD.**

(a) Hæmatoxylin, 1 grm.; Distilled Water, 300 cc.

(b) Chromate of Potassium, 1 grm.; Distilled Water, 200 cc. Small pieces which have been hardened in Alcohol or Picric Acid are placed in solution (a) for 12 to 24 hours; and then transferred for a similar time to solution (b). Wash the pieces thoroughly in Water, then dehydrate in Alcohol for embedding (saturation) in Paraffin. (*See p.* 17.)

This process is used for staining in bulk previous to serial cutting.

(x.) **BENDA'S COPPER HÆMATOXYLIN METHOD.**

For use after hardening in Chromic Acid or Flemming's Solution. Sections are placed for 24 hours in a 5 p. c. solution of Neutral Acetate of Copper, kept at about 40° C. (104° F.) washed thoroughly in Distilled Water; stained in a saturated aqueous solution of Hæmatoxylin, until the sections become dark grey or blackish. Decolourise to a rather light yellow with Hydrochloric Acid diluted 1 to 500 of Water. The sections must then be brought to a light blue colour by a second immersion in the copper solution. Wash in Water, dehydrate, clear and mount in Balsam.

CARMINE.

The principal uses to which Carmine is applied are—(1) Staining in bulk. (2) Staining sections of the nervous system. For the former, Grenacher's Alcoholic Borax Carmine is generally used, and for the latter Ammonia or Lithium Carmine, Grenacher's Alum Carmine, or Czoker's Alum Cochineal. (3) Staining Vegetable sections.

For sections other than the nervous system Lithium Carmine or Borax Carmine are useful stains.

Carmine under ordinary circumstances acts as a general stain, affecting the ground tissue as well as the nuclei. By subsequent treatment with Acidulated Alcohol or Acidulated Glycerine the colour is discharged from the ground tissue without seriously affecting the nuclei. Used in this way, Carmine becomes a good nuclear stain. It is also used extensively by some histologists in the form of Picro-carmine Solution.

Sections stained with the Carmines (except Alum Carmine) should not be washed in Water as the colour will be to a great extent removed, nor in Acidulated Water as the Carmine is precipitated on the sections. **Acidulated Alcohol** containing $\frac{1}{2}$ or 1 p. c. of Hydrochloric Acid (sp. gr. 1·16), in 70 p. c. Alcohol, or **Acidulated Glycerine** containing 1 p.c. of Glacial Acetic or of Formic Acid (sp. gr. 1·2), in equal parts of Glycerine and Water, should be employed. The former if Balsam be used for mounting, the latter if Farrant or Glycerine be employed.

STAINING IN BULK.

This method is used only in preparing material for serial cutting.

Tissues are placed in Grenacher's Alcoholic Borax Carmine from one to three days. They are transferred from the stain into $\frac{1}{2}$ p. c. Acidulated Alcohol till permeated by the acid, placed for a day in 90 p. c. Alcohol, and finally saturated with Absolute Alcohol before embedding.

NERVOUS SYSTEM.

Sections of spinal cord which has been hardened in Alcohol, are generally placed in Beale's Ammonia Carmine for two or three days, washed in 90 p. c. Alcohol, cleared, and mounted in Balsam. Much better staining however is obtained by using a stronger Carmine Solution, such as Lithium Carmine or an Ammonia Carmine five times the strength of Beale's Solution, *q.v.* Sections are stained in this for 24 hours and washed either in 90 p. c. Alcohol, or 70 p. c. Alcohol acidulated with $\frac{1}{10}$ p. c. Hydrochloric Acid. They may also be stained in either of the Alum Combinations as mentioned below.

When the central nervous system has been hardened in Chromates, a very prolonged immersion in Ammonia Carmine Solution is necessary, but in order to obviate this delay the following plan (due to Merkel) may be adopted:—

Sections are first placed in a solution of Chloride of Palladium (1 in 500) till they are of a straw colour (about 10—15 minutes); rinsed in Distilled Water; stained in the Ammonia Carmine Solution (where they quickly become an intense red), washed in 90 p. c. Alcohol, dehydrated, cleared, and mounted in Balsam. The medullary sheaths are stained yellow by the Palladium Solution, while the Carmine stains the neuroglia, axis cylinders, and ganglion cells a deep red.

Sections hardened either in Alcohol or the Chromates are advantageously stained in Alum Carmine or Alum Cochineal. Both of these solutions stain much more rapidly than Ammonia Carmine. The colour approaches to a violet, and is quite distinct from the red of Ammonia Carmine.

Either solution will stain Alcohol-hardened tissue in 30 minutes. Tissues hardened in the Chromates require 3 to 4 hours. When washed in 90 p. c. Alcohol (*without acid*), cleared and mounted in Balsam, good differential staining results. If, however, sections be left in the stain for 24 hours as recommended by some writers, overstaining takes place, and treatment with Acidulated Alcohol becomes necessary. This method also yields very good results, perhaps slightly better than the previous one, but occupies a much longer time.

FOR GENERAL USE.

Sections are stained with Lithium or Borax Carmine for 15 minutes to 2 hours according to the nature of the tissue and the depth of stain required, then treated with Acidulated Alcohol to decolourise the ground tissue, and so bring out the nuclei, washed in 90 p. c. Alcohol, cleared and mounted in Balsam.

As an example, a section of skin stained in Lithium Carmine for 50 minutes will be well differentiated by treatment with 1 p. c. Acidulated Alcohol for 10—15 minutes; when stained in Alcoholic Borax Carmine for 2 hours, acid washing for 4 minutes is sufficient for most purposes.

If Alum Carmine be used for the same length of time, treatment in Acidulated Alcohol is not required. The stained sections are first washed in water, then in 90 p. c. Alcohol, cleared and mounted in Balsam.

COUNTERSTAINING. *See Plasmatic Stains.*

(i.) **AMMONIA CARMINE.**

All Alkaline Solutions are good solvents for Carmine, and

a Solution in Ammonia is stated to have been introduced in 1858 by Gerlach, the formula being:—

Carmine, 1; Strong Liquid Ammonia, 1; Water, 100; the mixture exposed to the air for 24 hours to allow excess of Ammonia to evaporate, then filtered.

This solution easily undergoes change by the action of moulds and putrefactive ferments; but if time be given for these changes to run their course (some authors say "some months," others "two years") the solution, after filtration, keeps well and stains more precisely than a fresh solution. Beale's Solution containing Glycerine and Alcohol keeps much better than a simple Aqueous Solution, and answers every purpose. Frey's Solution is similar to Beale's. The Ammonia Carmine stain washes out completely in Water, also to some extent in 70 p. c. Alcohol, but not at all in 90 p. c. Alcohol.

(ii.) **BEALE'S AMMONIA CARMINE.**

Carmine, 10 grs.; Strong Solution of Ammonia, 30 mins.; Distilled Water, 2 oz.; Alcohol, ½ oz.; Pure Glycerine, 2 oz. Dissolve the Carmine in the Ammonia with the aid of heat; boil for a few seconds and let it cool; allow the excess of Ammonia to evaporate, and add the Water, Alcohol, and Glycerine; filter.

This solution will keep for months. Sometimes a little Carmine is deposited owing to escape of Ammonia, in which case 1 or 2 drops of Liquor Ammoniæ may be added to the 4 oz. of Carmine Solution.

(iii.) **STRONGER AMMONIA CARMINE.**

Carmine, 1 grm.; Strong Solution of Ammonia, 2 cc.; Distilled Water, 16 cc.; Alcohol, 8 cc.; Glycerine, 16 cc.; proceed as directed for Beale's Solution.

(iv.) GRENACHER'S ALUM CARMINE.

Carmine, 1 grm.; Alum (Ammonia), 5 grms.; Distilled Water, 100 cc. Dissolve the Alum in the Water, add the Carmine, and boil the solution for 20 minutes, allow it to cool, filter, and add Distilled Water to make the volume 100 cc.

This solution is chiefly used for the nervous system, but for ordinary staining it possesses the advantage that it is nuclear without after-treatment with acid.

(v.) CZOKER'S ALUM COCHINEAL.

Powdered Cochineal, 1 grm.; Alum, 1 grm.; Distilled Water, 100 cc.

Dissolve the Alum in the Water, add the Cochineal and boil; evaporate down to half of its original bulk, filter and add $\frac{1}{2}$ cc. of Liquid Carbolic Acid to the solution.

This is an excellent nuclear stain, and is chiefly used for the central nervous system after hardening in Chromium Salts. It is rather better for this purpose than Alum Carmine.

(vi.) GRENACHER'S BORAX CARMINE.

Carmine, 0·5 grm.; Borax, 2 grms.; Distilled Water, 100 cc. Heat the mixture to boiling point; then add a 5 p. c. solution of Acetic Acid, drop by drop, until the purple colour turns red; allow the mixture to stand for 24 hours, decant, filter, and add a few drops of Carbolic Acid to preserve the solution.

About 17 cc. of the Acetic Acid will be required.

(vii.) GRENACHER'S ALCOHOLIC BORAX CARMINE.

Carmine, 3 grms.; Borax, 4 grms.; Distilled Water 100 cc. Dissolve the Borax in the Water; add the Carmine and heat gently; then add 100 cc. of 70 p. c. Alcohol. Filter the solution (if necessary) before use. This is the best solution for staining in bulk. The pieces of tissue should be placed in the solution for 1 to 3 days, and then transferred to Acidulated Alcohol (p. 28).

Both of the above Borax Carmines are used for staining in bulk, but the Alcoholic solution is better adapted for the purpose, and keeps better than the aqueous which is apt to gelatinize after a time. It also makes a good stain for sections.

Although Acetic Acid is added to the Aqueous Solution, it does not remain uncombined, so that the staining has still to be supplemented by the usual Acid treatment.

(viii.) **ORTH'S LITHIUM CARMINE.**

Carmine, 2·5 grms.; Saturated Watery solution of Carbonate of Lithium, 100 cc. Digest the Carmine in the Lithium solution and filter.

Saturated Solution of Lithium Carbonate contains 1 grm. Carbonate of Lithium in 70 cc. of Distilled Water.

A very useful stain. Like other alkaline Carmine solutions, it stains both nuclei and ground substance, but is restricted to the nuclei by treating the sections for a short time with Acidulated Alcohol p. 28).

Lithium Carmine stains alcohol-hardened spinal cord very nicely and quickly. When the staining is deep, extraction with Acidulated 70 p. c. Alcohol gives a sharper definition than Acidulated 90 p. c. Alcohol.

PICRO-CARMINES.

Combinations of Carmine with Picric Acid have largely been used as effective double-stains. Various formulæ for these solutions exist, in which the Carmine is held in solution by Ammonia, Soda, or Lithia; but as the neutral picrates of these alkalies have in themselves no staining power, and the manner in which the Picric Acid is combined has never been accurately determined, no scientific explanation of their action can be given.

The following formula and method is that generally recommended, and in a large number of comparative experiments was found to give the best results.

(i.) **AMMONIA PICRO-CARMINE.**

Carmine, 1 grm.; Strong Solution of Ammonia, 3 cc.; Distilled Water, 5 cc. Dissolve the Carmine in the Ammonia and Water with a gentle heat, then add Saturated Aqueous Solution of Picric Acid, 200 cc.; heat to boiling and filter. This solution gives good results when used as follows:— Take a section which has been rinsed in Distilled Water and lay it out flat on a glass slide, drain off the superfluous water, then pour on to the section several drops of the Picro-Carmine Solution, warm the slide over a spirit lamp to a heat that can be borne by the hand when touched with the glass (if the section be too strongly heated it will shrivel), keep it about this temperature for 5 or 10 minutes, remove the excess of stain by tilting the glass and wiping it with a cloth or filter paper, leaving some of the stain in the section, then place one or two drops of Formic Farrant (p. 85) upon the section and apply the cover-glass. The staining of the section is much improved after it has been mounted 2 or 3 days and exposed to daylight.

A section of skin gives the most striking results by this method. Nuclei and the transverse muscular fibres stain red, the remainder yellow.

The following formula for Ranvier's Picro-Carmine, published in "Lee's Microtomists' Vade Mecum," 1890, p. 82, was given to Lee by Vignal as used in the Laboratory of Histology. of the College of France.

"Water, 1000; Picric Acid, 20; Carmine, 10; Ammonia, 50.

"Put them into a stoppered bottle, and leave them for 2 or 3 months in a warm place; then place the solution in a large crystallizing dish and allow it to putrefy. When the liquid has become reduced to four-fifths of its original bulk, remove the crystals that have formed at the bottom, dry them and dissolve them in a little Warm Water. Filter the solution, and examine it with the microscope to see if the Carmine is really dissolved; if not, add more Ammonia and Water, and repeat the above process. When the Carmine is so combined as to dissolve in the warm Water, then dissolve it, evaporate the solution to dryness in a stove, and reduce the Picro-Carminate of Ammonia to powder.

"For use dissolve 1 grm. of the powder in 100 cc. of Distilled Water."

(ii.) **SODA PICRO-CARMINE.**

Caustic Soda, 1 grm.; Distilled Water, 1000 cc.; dissolve and add Carmine, 10 grms.; boil, filter and make up the volume to 1000 cc. with Distilled Water; mix this with an equal volume of Distilled Water, and add 1 p. c. aqueous solution of Picric Acid so long as the turbidity, thus produced, continues to disappear on agitation.

(iii.) **PICRO-LITHIUM CARMINE.**

Lithium Carmine solution, 100 cc.; saturated aqueous solution of Picric Acid, 200—300 cc.

ANILINE NUCLEAR STAINS.

Fuchsine (Magenta), Gentian Violet, Dahlia Violet, Methylene Blue, Methyl Green, Hoffmann's Blue, Safranine, and Vesuvine are indirect nuclear stains.

The nuclei are differentiated from the ground substance by Alcohol washing.

A washing in Distilled Water, previous to the Alcohol, has the effect of fixing the stain in the nuclei so that it is less readily washed out in the Spirit. This is especially the case with Methylene Blue and Methyl Green, which *must* be washed thoroughly in water previous to being treated with Alcohol. If this be not done the colour will be almost completely removed by the Spirit. Treated in this way, however, the nuclear staining is very fine.

Vesuvine may or may not be previously washed in Distilled Water, according to the depth of stain required. If water be used the colour is stronger, but less easily removed from the ground tissue. The Violets and Safranine are better washed out at once in the Spirit without previous washing with water.

The differentiation in 90 p. c. Alcohol will vary from 10 to 30 minutes, according to the nature of the section.

When sections are to be also treated with one of the Plasmatic stains, the counterstaining will of course be done previous to the dehydration in Absolute Alcohol.

(i.) **METHYLENE BLUE.**

Methylene Blue, 0·25 grm.; Rectified Spirit, 20 cc.; Distilled Water, 80 cc.

To stain the nuclei, sections are placed in this solution for 5 minutes, rinsed thoroughly in Distilled Water, differentiated in 90 p. c. Alcohol, dehydrated, cleared and mounted in Balsam.

(ii.) **METHYL GREEN.**

Methyl Green, 0·5 grm.; Rectified Spirit, 20 cc.; Distilled Water, 80 cc.

To stain the nuclei, sections are placed in this solution for

5 minutes, rinsed thoroughly in Distilled Water, then differentiated in 90 p. c. Alcohol, dehydrated, cleared and mounted in Balsam.

The above solution, to which 1 p. c. of Glacial Acetic Acid has been added, is employed for fixing and staining nuclei in fresh vegetable tissues.

Methyl Green has been used for obtaining the peculiar pink reaction with amyloid tissue, but this is probably due to the fact that most methyl-greens contain methyl violet, as an impurity. Methyl green free from violet does not give the reaction.

(ii.a) **Ehrlich-Biondi Fluid.**

This is usually given as follows:—Saturated Aqueous Solution of Methyl Green, 5 cc.; Saturated Aqueous Solution of Orange, 10 cc.; Saturated Aqueous Solution of Acid Fuchsine, 2 cc.; for use dilute with about 40 volumes of Water.

As these concentrated solutions form a precipitate when mixed, it is better to dilute each of them with 20 to 40 volumes of water before mixing.

The solution is conveniently made as follows:—Methyl Green, 0·5 grm.; Distilled Water, 100 cc.: Acid Fuchsine, 0·5 grm.; Distilled Water, 40 cc.: Orange, 2 grms.; Distilled Water, 200 cc. Mix the solutions and filter before use.

In some cases (serial sections for instance) it is convenient to use all the dyes in one solution, but in other cases it is better to stain the nuclei with Methyl Green and counter-stain with Rubin and Orange.

Sections are usually placed in this stain for 12 hours, then washed, dehydrated, cleared and mounted.

This triple stain has also been strongly recommended as a counterstain to Hæmatoxylin, but in that case the Methyl Green is not required—Rubin and Orange (p. 42) answers equally well, if not better.

(ii.b) **Iodine Green.**

This is another name for Methyl Green. Formerly Methyl Iodide was employed in the manufacture of this green, and hence the names Methyl Green and Iodine Green. Now Methyl Bromide or Methyl Chloride is used in its place.

Iodine Green always contained large quantities of Violet, which was most probably the cause of the amyloid reaction assigned to this dye (*see Methyl Green*).

(iii.) **SAFRANINE.**

Safranine, 0·5 grm.; Rectified Spirit, 20 cc.; Distilled Water, 80 cc.

A good nuclear stain, and especially useful for tissues that are rapidly growing and dividing.

For nuclear staining, sections are placed in the solution for 5 to 10 minutes, differentiated in 90 p.c. Alcohol, dehydrated, cleared and mounted in Balsam.

(iv.) **VESUVINE (Bismarck Brown).**

Vesuvine, 0·5 grm.; Rectified Spirit, 20 cc.; Distilled Water, 80 cc.

A good nuclear stain, which is not easily washed out, and is permanent.

Sections are placed in this for 5 minutes, washed in 90 p.c. Alcohol with or without previous treatment with water, dehydrated, cleared and mounted in Balsam.

(v.) **GENTIAN VIOLET.**
DAHLIA VIOLET.

Gentian or Dahlia Violet, 0·5 grm.; Rectified Spirit, 20 cc.; Distilled Water, 80 cc.

To stain the nuclei, sections are placed in either of these Violet solutions for 5 minutes, then washed in 90 p.c. Alcohol, cleared and mounted in Balsam.

Gentian and Dahlia as well as Methyl Violet are generally included amongst the nuclear stains, but they are more generally used as Specific Stains. Gentian and Methyl Violets for micro-organisms; Methyl Violet for amyloid degeneration, and Dahlia Violet to demonstrate the "mastzellen."

Dahlia Violet has a strong red tint and Methyl Violet a strong blue tint, while Gentian Violet is intermediate between the two. Gentian and Dahlia are both very fine nuclear stains, but Methyl Violet washes out too readily in spirit to be of much use.

Gentian Violet may be made *directly* nuclear by adding 1 p. c. of Glacial Acetic Acid to the solution.

(vi.) **FUCHSINE (Magenta).**

Fuchsine, 1 grm.; Distilled Water, 150 cc.; Rectified Spirit, 50 cc.; dissolve and add Glycerine, 200 cc.

This fluid was devised by Dr. Ferrier for staining blood corpuscles. The sp. gr. is similar to that of Liquor Sanguinis. The coloured corpuscles of non-mammalian vertebrates alter but little in shape while they become stained (*Rutherford's Histology.* 1876).

(vii.) **HOFFMANN'S BLUE.**

Hoffmann's Blue, 1 grm.; Rectified Spirit, 20 cc.; Distilled Water, 80 cc.; Glacial Acetic Acid, 0·5 cc.

Immerse the sections for 10 minutes, rinse in Water, wash in 90 p.c. Alcohol, dehydrate, clear, and mount in Balsam.

PLASMATIC STAINS.

These colour the whole tissue uniformly. They are used as ground stains in connection with the nuclear and specific stains for the sake of contrast.

After the treatment for nuclear or specific staining, tissues if taken from Alcohol are placed for about a minute in Distilled

Water, then transferred to the Plasmatic stain and treated as directed.

After **Hæmatoxylin**, other Blues, and Green, use either Benzopurpurine (brownish red), Eosine (yellowish red), Erythrosine (pink), Orange, Rubin S., or Rubin and Orange; after **Carmine** and other Reds, use Picric Acid (yellow).

(i.) **BENZOPURPURINE.**

Benzopurpurine, 0·25 grm.; Rectified Spirit, 20 cc.; Distilled Water, 80 cc.

A good contrast stain to Hæmatoxylin and other blue nuclear stains. Sections are placed in this for 2 to 5 minutes, then washed in 90 p.c. Alcohol, dehydrated, cleared and mounted in Balsam.

(ii.) **EOSINE.**

Eosine (Water-soluble), 1 grm.; Rectified Spirit, 40 cc.; Distilled Water, 160 cc.

For counterstaining, the sections are placed in this solution for 2 to 5 minutes, then washed in 90 p.c. Alcohol, dehydrated, cleared and mounted in Balsam.

There are two classes of Eosines, characterised by their more ready solubility in water or in spirit. Each is useful according to the nature of the mounting media to be employed.

It is frequently stated that, when mounting in Balsam, the Eosine must be dissolved in the Alcohol used for dehydrating to prevent the stain washing out, but this is not at all necessary if the "water-soluble" variety be used.

It is a specific stain for red blood corpuscles, which it colours a copper red; also for the giant cells of Leprosy and Tubercle, staining them an orange-yellow.

(iii.) ERYTHROSINE.

Erythrosine, 1 grm.; Rectified Spirit 40 cc.; Distilled Water, 160 cc.

For counterstaining, sections are placed in this solution for 2 to 5 minutes, then washed in 90 p.c. Alcohol, dehydrated, cleared and mounted in Balsam.

(iv.) ORANGE.

Orange, 2 grms.; Rectified Spirit, 20 cc.; Distilled Water, 80 cc.

Sections are placed in this solution for 10 minutes, washed in 90 p.c. Alcohol, dehydrated, cleared and mounted in Balsam.

(v.) PICRIC ACID (ALCOHOLIC).

Picric Acid, 1 grm.; 70 p.c. Alcohol, 100 cc.

Principally used for staining fibrous tissue (skin, bone, &c.).

For counterstaining after Carmine omit the treatment with Acidulated Alcohol (p. 28), and rinse the sections for a couple of minutes in 70 p.c. Alcohol; transfer to Picric Acid about 5 minutes; dehydrate in Absolute Alcohol 3 or 4 minutes; clear in Cedar Oil; mount in Balsam.

It has been stated that Picric Acid must be added to the Alcohol used for dehydrating the sections, to prevent the stain washing out, but this is not necessary.

(vi.) ACID RUBIN.

Acid Fuchsine, 1 grm.; Rectified Spirit, 40 cc.; Distilled Water, 160 cc.

For counterstaining, sections are placed in this solution for 5 to 10 minutes, washed in 90 p.c. Alcohol, dehydrated, cleared and mounted in Balsam.

The dyes sold under the names Acid Rubin, Acid Fuchsine, and Acid Magenta have practically the same composition and give the same reactions, although different samples may vary somewhat in the shade of colour.

(vii.) **RUBIN AND ORANGE.**

Acid Fuchsine, 1 grm.; Orange, 6 grms.; Rectified Spirit, 60 cc.; Distilled Water, 240 cc.

For counterstaining, sections are placed in this for 5 to 10 minutes, washed in 90 p.c. Alcohol, dehydrated, cleared and mounted in Balsam.

When the elements of the ground tissue have the same affinity for each of the two colours, the result is a rich orange-red; where any selective affinity exists, either of the two colours will be more pronounced.

SPECIFIC STAINS.

These may be defined as stains which possess a peculiar affinity for certain elements in the tissues. The more important are Osmic Acid, Chloride of Gold, Nitrate of Silver, Victoria Blue, Dahlia Violet, Methyl Violet, Iodine, Safranine, Rose Bengale, Nigrosine, Acid Fuchsine, and Congo Red.

OSMIC ACID.

Aqueous solution, 1 p. c. (see also p. 8).

To be kept in an opaque bottle, and away from the light.

Useful for staining **fatty elements**, medullated nerve fibre, and elastic fibres. It is especially valuable for demonstrating fatty degeneration of organs.

Very small pieces of tissue in the fresh state are stained and fixed at the same time. It is also used in combination with Chromic Acid, as in Flemming's and Fol's solutions.

Material for sections should be hardened in Müller's Fluid, and cut on the freezing microtome. Alcohol partially dissolves the fat, and is, therefore, unsuited for the hardening, except when this is of no consequence. Osmic Acid also stains any fat which is left after treatment with Alcohol.

Sections are placed for 6 to 12 hours in $\frac{1}{6}$ to $\frac{1}{12}$ p. c. solution, carefully protected from the light, or for $\frac{1}{4}$ to $\frac{1}{2}$ hour in 1 p. c. solution; washed in Distilled Water, and mounted in Farrant or Glycerine.

CHLORIDE OF GOLD.

$\frac{1}{2}$ or 1 p. c. solution in Distilled Water.

Commercial Chloride of gold is not the pure Chloride $AuCl_3$, but the crystallized double Chloride of Gold and Sodium, containing 50 p. c. of Metallic Gold.

Commercial Chloride of Gold and Sodium is the above crystallized double Chloride mixed with an equal weight of Chloride of Sodium, and contains 25 p. c. of Metallic Gold.

Stains connective tissue corpuscles, and cartilage cells, but is principally used for tracing nerve endings. Can only be used for fresh tissues.

The methods used are very numerous and will be found well abstracted in Lee's Vade-mecum.

For fixed connective tissue corpuscles the following is the simplest.

(i.) A small piece of perfectly fresh tissue is placed in a ½ p. c. solution for half an hour, the solution being kept in the dark; the tissue becomes yellow; it is then washed in Distilled Water and exposed to daylight in Distilled Water, acidulated slightly with Acetic Acid. In a couple of days the tissue becomes purplish or violet brown, and must be mounted in Glycerine.

(iii.) For tracing nerve-fibres the process known as **Ranvier's Lemon-Juice Method** is generally applicable.

The fresh tissue is soaked in Lemon juice for 5 to 10 minutes (40 grains of Citric Acid in an ounce of water forms a convenient substitute); it is then rapidly washed in Distilled Water and transferred to a 1 p. c. Gold Chloride solution for from 10 minutes to 1 hour, according to the nature of the tissue. After washing in Water it is placed in 50 cc. of Water containing 2 drops of Acetic Acid and exposed to the light, or—if we do not wish to retain the superficial epithelium—after treating with Lemon Juice and Gold, the tissue is placed for 24 hours in Formic Acid (Sp. gr. 1·2) diluted with three times its volume of Water, and kept in the dark.

NITRATE OF SILVER.

½ p. c. solution in Distilled Water.

Used for staining the tesselated epithelium of serous membranes, lymphatics and blood-vessels, the matrix of connective tissue and cartilage, cancerous growths, and the eye; it also hardens tissue.

The tissue must be fresh and the sections very thin. They are first washed in Distilled Water to remove the Chlorides, and are then placed in the Silver solution for half an hour in the dark. After this they are washed in Distilled Water, placed in ordinary Water, and exposed to diffuse light till brown; they must be mounted in Glycerine or Glycerine Jelly, and kept in the dark.

"Fixing" with Hyposulphite of Sodium has been suggested, to render the specimens permanent in light.

VICTORIA BLUE.

Victoria Blue, 0·25 grm.; Rectified Spirit, 20 cc.; Distilled Water, 80 cc.

Stain the sections 3 to 5 minutes, rinse in Distilled Water about 2 minutes, wash in 90 p. c. Alcohol 3 or 4 minutes, dehydrate in Absolute Alcohol 3 or 4 minutes, clear in Cedar Oil, and mount in Balsam.

This is especially useful for demonstrating elastic tissue, staining the elastic fibres an intense blue; it is, however, somewhat readily extracted by Alcohol.

It stains best after hardening in Chromium Salts, but this is not necessary. It also acts as a nuclear stain.

DAHLIA VIOLET.

Dahlia Violet, 2 grms.; Rectified Spirit, 25 cc.; Distilled Water, 70 cc.; Glacial Acetic Acid, 5 cc.

Specially useful for demonstrating the granules in Ehrlich's **Mastzellen.** The tissue must be hardened in Alcohol, as is the case with bacteria. Stain for some hours, then wash out in Alcohol until nearly colourless, dehydrate, clear in Cedar Oil, and mount in Balsam.

The uniformly round granules of these plasma cells (granular cells) are stained by Aniline dyes (especially the Violets) in a similar manner to bacteria, and are chiefly distinguished from the latter by the fact that the granules are always grouped in the form of cell-like structures, and the Aniline stain is removed from the granules, but not from bacteria, by treatment with a weak solution of Carbonate of Potassium.

STAINS FOR AMYLOID DEGENERATION.
METHYL VIOLET.

Methyl Violet, 0·5 grm.; Water, 200 cc.; Glacial Acetic Acid, 5 c.c.

Stain sections for 20 minutes; wash thoroughly in Distilled Water, then in a mixture of equal parts Glycerine and Water so long as any colour is taken out (generally about 2 hours) and finally mount in Farrant.

The amyloid material is stained pink and the normal tissue a bluish-green colour.

When stained as usual without the addition of Acetic Acid, the colour-contrast is not so striking. It may, however, be improved by after-treatment with dilute acid, but it is better to combine the acid with the stain.

Most Methyl and Iodine Greens give the same reaction owing to the quantity of Methyl Violet which they contain.

IODINE.

Iodine, 1 grm.; Iodide of Potassium, 2 grms.; Distilled Water, 300 cc.

Solution is most easily effected by using 10 cc. of the Water to dissolve the solids, then diluting with the remainder of the water.

This is the same as Gram's solution for Micro-organisms.
An alternative formula is:—
Liquor Iodi, B.P. 10 cc.; Distilled Water, 140 cc.

Stain the sections for 5 minutes; wash in Water and mount in thick Farrant to which a fourth of its volume of Liquor Iodi has been added.

Under the microscope (a low power answers best) by transmitted light, the section appears coloured various shades of yellow, but by reflected light the amyloid substance appears almost black while the normal tissue remains unaltered.

SAFRANINE.

Safranine, 1 grm.; Rectified Spirit, 20 cc.; Distilled Water, 80 cc.

Stain the sections for 2 minutes and wash in water. The amyloid substance gives a bright orange-red colour contrasting well with the normal pink.

If to be mounted, wash the sections in a mixture of equal parts Glycerine and Water, frequently changed until no more colour is removed (12 hours is generally sufficient), then in pure Glycerine in the same way, and finally mount in Farrant.

By passing rapidly through 90 p.c. Alcohol and Absolute Alcohol into Cedar Oil, the section may be mounted in Balsam, but the bright orange colour of the Amyloid is always impaired by contact with spirit.

ROSE BENGALE.

Rose Bengale, 1 grm.; Rectified Spirit, 20 cc.; Distilled Water, 80 cc.

Stain the sections for five minutes, wash in 90 p.c. Alcohol, dehydrate, clear and mount in Balsam.

It demonstrates very well the commencement of **amyloid**

degeneration, staining it a bright red round the blood vessels. With old amyloid the results are good, but not so striking.

This was noticed for the first time by Warwick in my laboratory, when using Rose Bengale in some experiments on ground stains for different tissues.

Used as above it is an excellent stain for elastic fibres.

STAINS AND METHODS FOR NERVE CENTRES.
NIGROSINE. Aniline Black. Aniline Blue Black.

Nigrosine, 2 grms.; Water, 100 cc.

Particularly useful for staining unhardened sections of brain; but it will stain equally well sections of spinal cord which have been hardened in Alcohol.

Sections of hardened material are placed in this for 10 to 15 minutes, washed in 90 p.c. Alcohol, cleared and mounted in Balsam.

ACID FUCHSINE.

Acid Fuchsine, 2 grms.; Water, 100 cc.

Sections of hardened material are placed in this solution for 2 or 3 minutes, washed in 90 p. c. Alcohol, cleared and mounted in Balsam.

CONGO RED.

Congo Red, 2 grms.; Water, 100 cc.

Sections are placed in this solution for 2 or 3 minutes, washed in 90 p.c. Alcohol, cleared and mounted in Balsam.

This dye, but little known, is one of the best stains which I have tried for the central nervous system.

WEIGERT'S METHOD.

The brain and spinal cord must be hardened in one of the Bichromate Solutions, and completed in Alcohol.

For embedding in Celloidin or Gum, see p. 15.

SPECIFIC METHODS.

The pieces embedded in Celloidin are taken from the spirit and placed for 1 or 2 days in a Saturated Aqueous Solution of Acetate of Copper diluted with an equal bulk of water, kept at about 40° C. (104° F.), and then transferred to 80 p.c. Alcohol till required for cutting: or the sections can be cut first, and afterwards immersed in the Copper Solution.

The sections are well washed in 90 p.c. Alcohol and transferred to Weigert's Hæmatoxylin (p. 26) for a few hours to 2 days, according to the differentiation required, they should be opaque, and a deep blue-black; and then well washed for 2 or 3 days in Distilled Water.

They are now decolourised for $\frac{1}{2}$ to 2 hours in the following solution:—

Borax, 2 grms.; Ferricyanide of Potassium, 2·5 grms.; Water, 200 cc.

While in this solution the sections must be carefully watched, and removed as soon as the grey and white substances are sharply defined.

They are again washed in Water for half an hour, dehydrated, cleared, and mounted in Balsam.

The white substance of Schwann is coloured a purple black, the connective tissue, axis cylinders, ganglion cells, and degenerated tracts in cases of sclerosis, an orange brown.

PAL'S METHOD.

A modification of Weigert's. Takes less time, and admits of counter-staining.

The sections after being cut are placed in the following solution:—

Hæmatoxylin, 0·75 grms.; Distilled Water, 90 cc.; Absolute Alcohol, 10 cc.—to which is added, just before use, a saturated solution of Lithium Carbonate in the proportion of

3 drops to every 10 cc. of the Hæmatoxylin solution. This is practically Weigert's Hæmatoxylin.

In this stain the sections are allowed to remain from 5 to 6 hours until they are opaque, and have a dark bluish-black colour.

They are then removed to Distilled Water and thoroughly washed till no more colour comes away. If the sections are not sufficiently stained a little Carbonate of Lithium solution is added to the Water.

The sections are now placed in 0·25 aqueous solution of Permanganate of Potassium from 15 to 20 seconds to differentiate; they are then placed in—

Pal's Solution.—Oxalic Acid, 1 grm.; Potassium Sulphite, 1 grm.; Distilled Water, 200 cc.; and allowed to remain till the white and grey matters are distinctly defined. This takes from 1 to 2 minutes; should black spots appear, the sections must be replaced in the Permanganate of Potassium solution, and put back into Pal's solution; they are then washed for 15 minutes in Water.

By this means the medullary sheaths are stained a bluish-black colour, while the rest of the tissue remains unstained.

The nuclei may now be stained in Alum Carmine.

Finally dehydrate, clear, and mount.

PAL-EXNER METHOD.

This is the most rapid of all, and takes about 3 days both to harden and mount.

The spinal cord or brain is cut into $\frac{1}{4}$-in. cubes and immersed in ten times its bulk of $\frac{1}{2}$ p.c. solution of Osmic Acid for 2 days, the solution being changed each day; the pieces are then washed in Water, transferred to Absolute Alcohol, and embedded in Celloidin or Paraffin.

As the sections are cut they are placed in Glycerine, then

SPECIFIC METHODS.

washed in Water, treated in the Permanganate of Potassium and Pal's solution, and counter-stained with Carmine; dehydrated, cleared, and mounted in Balsam.

GOLGI'S SUBLIMATE METHOD.

Small pieces (about $\frac{1}{2}$ in. cubes) of the tissue are hardened (15 to 30 days) in about 200 cc. of Muller's Fluid (p. 4) which should be frequently changed.

They are then transferred to aqueous solution of Corrosive Sublimate (0·25 to 1 p.c.) which must be changed from time to time as it becomes coloured by the Bichromate.

The sublimate treatment takes from 8 to 10 days or even longer.

Sections of the material thus treated are well washed with water.

Subsequent treatment with a weak solution of Sulphide of Sodium darkens the stain and makes it sharper.

STAINING CELLULOSE AND ITS MODIFICATIONS.

Unaltered Cellulose.—This is stained by Carmine, Hæmatoxylin, and most of the aniline dyes.

It is coloured blue with Chlorzinc Iodine (Schulze's Solution); also by treatment with Iodine followed by Sulphuric Acid diluted 1 to 3 of water.

It is dissolved by Sulphuric Acid (undiluted).

Staining with Carmine.—Place a section in Grenacher's Alcoholic Borax Carmine (p. 32) for 15 to 30 minutes, rinse for a second in Distilled Water, transfer to 90 p. c. Alcohol; dehydrate, clear, and mount in Balsam.

The Carmine washes out quickly in Water, therefore the section must not be left there any length of time, but it is better to just rinse the sections in Water, than to transfer them direct from the Carmine solution to the Alcohol.

Cellulose, nuclei, and protoplasm in general are stained.

For staining nuclei only, see "Nuclear Staining," p. 20.

Staining with Hæmatoxylin.—Rinse the sections in Distilled Water for a minute or two, place them for 5 minutes in Ehrlich's or Delafield's Solution (p. 24) diluted 1 to 9 of Distilled Water; rinse in Distilled Water and transfer to Tap Water (if alkaline) or to a weak solution of Bicarbonate of Sodium (1 grain to 2 ounces), until the staining acquires a blue tint; dehydrate, clear, and mount in Balsam.

Staining with Aniline Dyes.—Methyl Green, Methylene Blue, Safranine and Fuchsine give good results. Place sections in ¼ p. c. solutions of either of these for 3 or 4 minutes, rinse in Water, dehydrate, clear, and mount.

Lignified Cellulose.—When the cellulose undergoes the change known as lignification, its reactions are altered. It is coloured (i.) yellow or yellowish-brown by Chlorzinc Iodine; (ii.) red with Phloroglucin followed by Hydrochloric Acid; (iii.) yellow with Aniline Chloride. The results with Phloroglucin are more definite and reliable than with Chloride of Aniline.

It is generally stated that Hæmatoxylin and Carmine do *not* stain lignified tissue. With Carmine this is practically true; but while it is probable that pure "Lignin" is unaffected by Hæmatoxylin, the tissue generally takes on a modified stain, which is lighter in proportion to the extent of the lignification. In young stems it is sometimes noticed that Hæmatoxylin has a special staining power for the "middle lamella" both of wood and bast, which Carmine, on the other hand, does not affect at all.

The Aniline Dyes mentioned under "Cellulose Staining" will also stain equally well lignified tissues.

It is stained by Picric Acid, but the dye washes out easily in Water or Alcohol. Good results however can be obtained by carefu treatment.

DOUBLE STAINING MIXED TISSUES.

The most striking results are obtained by using Carmine and Methyl Green. The staining with Carmine and Picric Acid is also very effective, but as the Picric washes out very rapidly, the sections should not remain in the Alcohol longer than is required to dehydrate them. A thick section takes longer to dehydrate, but holds the stain well; a thin section dehydrates quickly, but also rapidly loses the stain.

The differentiation with Aniline Blue and Picric Acid is not so sharp as the above, but more delicate gradations are obtained, from blue through green to yellow.

The general distribution of the lignified tissue is best observed with a low power, say an inch object glass, when the different colours should show distinctly.

With Carmine and Methyl Green.—Sections of stems and roots containing cellulose and lignified tissue are rinsed in Distilled Water; place them in Methyl Green Solution for 3 or 4 minutes; rinse in Distilled Water; wash in 90 p. c. Alcohol for 5 or 10 minutes, which to a great extent removes the Green from the cellulose portion, and thus obtains a purer red with the Carmine; place the sections in Grenacher's Alcoholic Borax Carmine for 15 or 20 minutes; rinse quickly in Distilled Water to remove adhering stain; pass through 90 p. c. Alcohol, dehydrate, clear, and mount in Balsam.

Acid Green is frequently employed in the place of Methyl Green, but if the latter is free from Violet, or nearly so, it is the better of the two.

With Carmine and Picric Acid.—Rinse the sections in Distilled Water; place them in Grenacher's Alcoholic Borax Carmine 15 to 20 minutes; rinse for a second in Distilled Water; transfer to Picric Acid (Alcoholic) for 5 or 10 minutes; dehydrate in Absolute Alcohol for 3 to 5 minutes, clear in Cedar Oil, and mount in Balsam.

With Aniline Blue and Picric Acid.—Rinse the sections in Water; place them in Hoffmann's Blue (p. 56) for 10 minutes; transfer to Picric Acid (Alcoholic) for 10 minutes; dehydrate in Absolute Alcohol for 3 or 4 minutes, clear in Cedar Oil, and mount in Balsam.

This method shows very well the bordered pits in a vertical section of pine stem.

CHLOR-ZINC IODINE (Schulze's Solution).

No good working formula has been published for this reagent: the points being to obtain a clear concentrated solution which does not crystallize on keeping, and which dissolves sufficient Iodine.

The following will yield a permanent solution, which gives good results, staining unaltered cellulose a violet-blue and lignified tissue yellow. Starch is of course stained blue.

Evaporate 100 cc. of Liquor Zinci Chloridi (B.P.) to 70 cc.; dissolve in it 10 grms. of Iodide of Potassium; then add 0·2 grm. Iodine: shake at intervals till saturated.

PHLOROGLUCIN.

Phloroglucin, 1 grm.; Rectified Spirit, 20 cc.; Distilled Water, 80 cc.

Sections containing lignified tissue placed in this solution for about 15 minutes, then treated with strong Hydrochloric Acid, are stained a cherry-red colour. The depth and distribution of the colour are in proportion to the extent of the lignification.

CHLORIDE OF ANILINE.

Chloride of Aniline, 2 grms.; Rectified Spirit, 65 cc.; Distilled Water, 35 cc.; Strong Hydrochloric Acid, 2 cc.

Sections containing lignified tissue placed in this solution for 15 to 30 minutes are stained a yellow colour, but the differentiation is not so good as with Phloroglucin.

METHYL GREEN.

Methyl Green, 0·25 grm.; Rectified Spirit, 20 cc.; Distilled Water, 80 cc.

PICRIC ACID (Alcoholic).

Picric Acid, 1 grm.; 70 p. c. Alcohol, 100 cc.

STAINING SIEVE AREAS.

These are best stained by Hoffmann's Blue, but Eosine is frequently employed for this purpose. They can also be demonstrated by Hæmatoxylin, using the process given above for cellulose.

HOFFMANN'S BLUE.

Hoffmann's Blue, 1 grm.; Rectified Spirit, 20 cc.; Distilled Water, 80 cc.; Glacial Acetic Acid, 0·5 cc.

Place the sections in the stain for 5 to 10 minutes, rinse in Distilled Water in which the colour is partially removed, mount in Glycerine; or dehydrate, clear, and mount in Balsam.

This is also a very good nuclear stain.

EOSINE.

Eosine, 2 grms; Rectified Spirit, 20 cc.; Distilled Water, 80 cc. Place the sections in the stain for 5 to 10 minutes, rinse in Water, wash in 90 p. c. Alcohol, dehydrate, clear, and mount in Balsam.

MICRO-ORGANISMS.

STAINS.

In the investigation of micro-organisms the following solutions of the Aniline Stains and of the Intermediate Reagents are employed.

Different samples of the same aniline dye will differ in their behaviour to solvents. Some will make a clear solution, others will leave more or less insoluble residue. The solutions should therefore be filtered or decanted. It is safer to filter aqueous solutions each time before use.

Aqueous Solutions of the dyes do not keep well, therefore Alcohol has been added as a preservative.

(i.) **VESUVINE (Bismarck Brown).**

(*a*) **Glycerine Solution.**—Saturated solution in equal parts of Glycerine and Distilled Water.

(*b*) **Aqueous Solution.**—Vesuvine, 1 grm.; Rectified Spirit, 20 cc.; Water, 80 cc.

(ii.) **METHYLENE BLUE.**

(*a*) **Saturated Alcoholic Solution.**

Methylene Blue, 15 grms.; Absolute Alcohol, 100 cc.; shake at intervals for one or two days until saturated, then filter.

Only a small percentage of the dye is soluble in Alcohol, yet the depth of colour obtainable in the solution increases in proportion to the quantity added up to 15 p. c., but not further.

(*b*) **Aqueous Solution.**—Methylene Blue, 1 grm.; Rectified Spirit, 20 cc.; Distilled Water, 80 cc.

A freshly made 2 p. c. solution is 50 p. c. stronger in colour; but it crystallizes out in cold weather. Owing to the influence of some soluble impurity, a 10 p. c. solution gives a weaker colour than a 2 p. c.

(c) **Koch's Methylene Blue Solution.**—Saturated Alcoholic Solution of Methylene Blue, 1 cc.; 10 p. c. solution of Caustic Potash, 0·2 cc.; Distilled Water, 200 cc.

This formula is taken from Koch's paper, "Die Aetiologie der Tuberculose in Berliner Klinische Wochenschrift," 1882, No. 15, p. 221. In most English text-books, the quantity of alkali is increased ten times, owing to the omission of a decimal point in the translation from which these are copied.

(d) **Löffler's Solution.**—Concentrated Alcoholic Solution of Methylene Blue, 30 cc.; Solution of Caustic Potash (1 in 10,000), 100 cc.

(e) **Kühne's Carbolic Methylene Blue.**— Methylene Blue, 1·5 grms.; Absolute Alcohol, 10 cc. Rub in a mortar, and add 100 cc. of a 5 p. c. Aqueous Solution of Carbolic Acid.

(f) **Kühne's Aniline Oil Solution.**—Rub up in a mortar as much Methylene Blue as will go upon the point of a knife, with 10 cc. of Aniline, and pour the whole unfiltered into a bottle; after a time the undissolved pigment will settle at the bottom.

(iii.) **GENTIAN VIOLET.**

(a) **Concentrated Alcoholic Solution.**—Gentian Violet, 25 grms.; Absolute Alcohol, 100 cc.

(b) **Aqueous Solution.**—Gentian Violet, 1 grm.; Rectified Spirit, 20 cc.; Distilled Water, 80 cc.

(c) *****Gentian-Violet Aniline-Water.**—Aniline Water, 100 cc.; Concentrated Alcoholic Solution of Gentian Violet, 11 cc.; Absolute Alcohol, 10 cc.

(iv.) **METHYL VIOLET.**

(a) **Concentrated Alcoholic Solution.**—Methyl Violet, 25 grms.; Absolute Alcohol, 100 cc.

* " Erhlich-Weigert-Koch " formula.

(*b*) **Aqueous Solution.**—Methyl Violet, 1 grm.; Rectified Spirit, 20 cc.; Distilled Water, 80 cc.

(*c*) **Kühne's Solution.**—Methyl Violet, 1 grm.; Distilled Water, 90 cc.; Alcohol, 100 cc.

(*d*) ***Methyl-Violet Aniline-Water.**— Aniline Water, 100 cc.; Concentrated Alcoholic Solution of Methyl Violet, 11 cc.; Absolute Alcohol, 10 cc.

VICTORIA BLUE.

Victoria Blue, 2 grms.; Alcohol (50 p. c.), 100 cc.

(vi.) **FUCHSINE (Magenta).**

(*a*) **Concentrated Alcoholic Solution.**— Fuchsine, 25 grms.; Absolute Alcohol, 100 cc.

(*b*) **Aqueous Solution.**—Fuchsine, 1 grm.; Rectified Spirit, 20 cc.; Distilled Water, 80 cc.

(*c*) ***Fuchsine Aniline-Water.**—Aniline Water, 100 cc.; Concentrated Alcoholic Solution of Fuchsine, 11 cc.; Absolute Alcohol, 10 cc.

(*d*) **Ziehl-Neelsen Solution, or Kühne's Carbolic Fuchsine.**—Fuchsine, 1 grm.; Absolute Alcohol, 10 cc. Dissolve and add 100 cc. of a 5 p. c. Aqueous Solution of Carbolic Acid.

(*e*) **Magenta Solution (Gibbes').**—Magenta, 2 grms.; Aniline Oil, 3 grms.; Rectified Spirit, 20 cc.; Distilled Water, 20 cc.

(*f*) **Gibbes' Double Stain.**—Magenta, 2 grms.; Methylene Blue, 1 grm. Well triturate the mixture, and then add slowly Aniline Oil, 3 cc., dissolved in Rectified Spirit, 15 cc.; subsequently add Distilled Water, 15 cc., and keep stain in a stoppered bottle.

* "Erhlich-Weigert-Koch" formula.

(vii.) **SAFRANINE.**

Kühne's Aniline Oil Solution.—Made in the same way as that of Methylene Blue.

Babes' Solution.—Saturate Aniline Water with Safranine at 60° C. (140° F.), and filter.

(viii.) **METHYL GREEN.**

Aniline Oil Solution.—Made in the same way as Methylene Blue.

(ix.) **BLACK BROWN.**

Carbolic Black Brown.—Made in the same way as Carbolic Fuchsine.

(x.) **ORSEILLE (Wedl).**

Glacial Acetic Acid, 5 cc.; Absolute Alcohol, 20 cc.; Distilled Water, 40 cc.: mix and add Orseille (Archil, from which excess of Ammonia has been driven off), to form a dark reddish fluid.

Israel has more recently used a strong solution of Orcein in Acetic Acid.

INTERMEDIATE REAGENTS.

(i.) GRAM'S SOLUTION.

Iodine, 1 grm.; Iodide of Potassium, 2 grms.; Distilled Water, 300 grms.

(ii.) ANILINE WATER.

Distilled Water, 100 cc.; Aniline, 5 cc. To be well shaken and the emulsion filtered twice.

This is the formula generally given, but as the solubility of Aniline is 1 in 27 of Distilled Water, a clear solution of practically the same strength can be obtained by mixing 3 cc. of Aniline with 90 cc. of Distilled Water; it can also be made extemporaneously by pouring a few drops of Aniline into a test-tube and adding Distilled Water gradually (with agitation) until the turbidity disappears. Aniline Water soon spoils, and therefore should be made fresh as required. With the addition of 10 p. c. of Alcohol it keeps much better.

(iii.) WEAK ACIDULATED WATER.

Hydrochloric Acid, 1 cc.; Distilled Water, 1000 cc.

(iv.) LITHIA WATER.

Saturated Aqueous solution of Lithium Carbonate (1 in 70), 1 cc.; Water, 30 cc.

(v.) **SULPHURIC ACID AND ALCOHOL.**

Sulphuric Acid (sp. gr. 1·84), 10 cc.; Alcohol (90 p. c.), 90 cc.

(vi.) **SULPHURIC ACID.**

Sulphuric Acid (sp. gr. 1·84), 10 cc.; Distilled Water, 30 cc.

(vii.) **NITRIC ACID.**

Pure Nitric Acid (sp. gr. 1·42) 10 cc.; Distilled Water, 30 cc.

(viii.) **SULPHANILIC AND NITRIC ACID.**

Sulphanilic Acid (saturated aqueous solution), 30 cc.; Nitric Acid (sp. gr. 1·42), 10 cc.

(ix.) **DILUTED ACETINE (CHLORHYDRIN) BLUE.**

Acetine Blue (Chlorhydrin Blue), 10 grms.; Absolute Alcohol, 10 cc.; Distilled Water, 90 cc.

Chlorhydrin Blue mentioned by Kühne is not obtainable commercially. Acetine Blue appears to be the same article.

(x.) **FLUORESCINE.**

Alcoholic Solution.—Yellow Fluorescine, 1 grm., rubbed in a mortar with 50 cc. of Absolute Alcohol; the whole poured into a bottle and allowed to settle.

GENERAL METHODS.

WITH ANILINE VIOLETS.

It would appear that the first application of Aniline colours to Bacteriological investigation was due to **Weigert**, who stained in a simple **aqueous** solution of Methyl Violet.

To better fix the colouring matter, **Erhlich** used a saturated solution of **Aniline in Water** as a mordant for the dye, the dye being dissolved in this, or added in the form of a concentrated alcoholic solution till a slight opacity appears.

Weigert subsequently proposed definite volumes of the solutions which have been generally adopted, with the further addition (by Koch) of 10 per cent. of Absolute Alcohol to make a more permanent solution. Koch states that the solution so made will remain good for 10 days, without requiring to be filtered each time of using. Weigert later substituted Gentian Violet for Methyl Violet, as the former stains more quickly and does not wash out so readily in Alcohol.

Koch produces an isolated staining of bacteria in sections, by staining them in aqueous solution of Methyl Violet, Fuchsine, or Methylene Blue, and washing them in a saturated aqueous solution of Carbonate of Potassium diluted with an equal bulk of water; this removes the colour from the nuclei, leaving the bacteria stained: the sections are then dehydrated, cleared and mounted.

The general methods now adopted are:—

For Living Bacteria.

The fluid under examination is mixed on the slide with a drop of an aqueous (free from Alcohol) solution of Gentian Violet, Fuchsine, or Methylene Blue.

For Tissues and Cover-glass Preparations.

Stain for a few minutes in the Gentian-Violet Aniline-Water (p. 58); wash in water, and all but decolourise the ground substance in strong Alcohol; counterstain, if required, and mount as usual.

Gram's Method.

Sections are removed from weak Alcohol, and placed in Gentian-Violet Aniline-Water for 1 to 3 minutes. (Tubercle sections 12 to 24 hours.)

The sections are quickly rinsed in Absolute Alcohol and transferred into Gram's solution (p. 61), until they acquire a brown colour; this takes about 1 to 3 minutes. The sections are then washed in 90 p. c. Alcohol until they are a pale yellow, dehydrated, cleared, and mounted in Balsam.

The sections are frequently counterstained with Eosine or Vesuvine.

Weigert's Modification of Gram's Method.

As the prolonged washing in Alcohol may remove colour from the organisms, Weigert proposed the substitution of Aniline for Alcohol, and conducts the process on a slide. Place the section on a glass slide, and stain it with a few drops of Gentian-Violet Aniline-Water; remove the excess of fluid, and apply for 2 minutes a few drops of Gram's solution. Remove this by gently blotting it off; then wash the section by allowing Aniline to flow gently backwards and forwards over it; when no more colour comes away, pour off the Aniline, and treat in a similar way with Xylol for about a minute, and mount in Balsam.

Previous to this process for bacteria, the nuclei in the sections may be stained with Carmine (p. 28).

Kühne's Modification of Gram's Method.

Kühne uses Fluorescine Alcohol after Gram, in the place of Alcohol for washing, and adopts Weigert's Aniline for clearing; he also stains the nuclei with Carmine (nuclear staining, p. 28), previous to commencing the following process.

Sections are stained for 5 minutes in the solution of Methyl Violet, diluted one-sixth with a 1 p. c. aqueous solution of Ammonium Carbonate, or in the Victoria Blue solution; they are then rinsed thoroughly in Water, and transferred to Gram's solution from 2 to 3 minutes, again rinsed in water, and the excess of stain extracted in Fluorescine Alcohol. They are then passed through pure Alcohol, Aniline, Ethereal Oil, and Xylol, and mounted in Balsam.

WITH METHYLENE BLUE.

Löffler's Universal Staining Fluid.

Concentrated Alcoholic solution of Methylene Blue, 30 cc.; Solution of Caustic Potash (1 in 10,000 of water), 100 cc. This mixture will not keep good for many days, and should be filtered each day before use.

Löffler found that most bacteria stained better in this solution than in the weaker solution used by Koch for the Tubercle bacillus.

Sections are placed in this solution for a few minutes (Tubercle sections for some hours); remove excess of stain by immersion in very weak Acetic Acid ($\frac{1}{2}$ p. c.) for a few seconds, dehydrate in Absolute Alcohol, clear in Cedar Oil and mount in Balsam.

Kühne's Methods.

The sections are transferred from Alcohol to Carbolic Methylene Blue (p. 58), and allowed to remain in this for half an hour (leprosy bacilli require longer — 2 hours). The sections are next rinsed in Water, and then placed in Weak Acidulated Water until they are of a pale blue colour (this must be done carefully or too much colour will be removed); they are then rinsed in Lithia Water (p. 61), and transferred to a basin of pure Water. The sections are taken up one by one upon the point of a glass needle and dipped into Absolute Alcohol in which some Methylene Blue has been dissolved. They are then transferred to Methylene Blue Aniline Oil to dehydrate, rinsed in Aniline and then placed for about a couple of minutes into some ethereal oil of low density, such as Terebene, to clear, then into Xylol, and mounted in Balsam.

The above method is one that will apply to the staining of all micro-organisms, with the exception of the bacilli of leprosy and mouse septicæmia; these it stains insufficiently.

In order to show satisfactorily the structure of the tissue, Kühne recommends the following plan :—

After staining in Carbolic Methylene Blue, the sections are decolourised in a diluted solution of Chlorhydrin Blue (p. 62), which requires from 10 to 60 minutes. The sections are then passed through Alcohol, Aniline, Ethereal Oil, and Xylol, and, in order to double-stain, the sections are taken from the Xylol and placed in Safranine Aniline Oil, diluted to about 4 or 5 times its bulk with Aniline. After remaining in this from 2 to 10 minutes they are passed again through Ethereal Oil and Xylol before mounting.

WITH FUCHSINE.

Ehrlich's Method, *see Tubercle Bacilli,* p. 68.

Kühne's Methods.

Method of Staining Bacteria not belonging to the Group of Tubercle Bacilli in the Tissues with Fuchsine (Kühne).

The bacilli of **mouse septicæmia, anthrax,** and other bacilli, and several kinds of **cocci** may be stained by the following method, which will not apply to the bacilli of glanders and typhoid fever:—

Sections are first dehydrated in Alcohol and then stained from 3 to 5 minutes in Carbolic Fuchsine, rinsed in water, dipped for 1 minute into Alcohol, and extracted in Methyl Green Aniline Oil, which requires from 15 minutes to 2 hours, according to the thickness of the sections. The sections are then passed through Ethereal Oil and Xylol.

Method of Staining with Black Brown and Fuchsine (Kühne).

Anthrax sections are stained for 5 minutes in Carbolic Black Brown, rinsed in Lithia Water (p. 61), and then in 90 p. c. Alcohol; the sections are placed for 5 minutes in Carbolic Fuchsine, and decolourised in Fluorescine Alcohol.

In this method Carbolic Black Brown acts as a mordant to Fuchsine, so that the Anthrax bacilli do not lose the red colour in the Fluorescine Alcohol.

SPECIAL METHODS.

TUBERCLE BACILLUS.

Koch's Original Method.

Sections or cover-glass preparations are placed in Koch's Methylene Blue Solution (p. 58) and left in it for 20 to 24 hours, or $\frac{1}{2}$ to 1 hour if warmed to 40°C. (104°F.); they are then placed in aqueous solution of Vesuvine for 2 minutes; this removes the blue colour from the tissue, staining it brown, while the bacilli remain blue; the excess of colour is washed out in Distilled Water; if cover-glass preparations, they are then dried and mounted in Balsam; if sections, they are dehydrated in Absolute Alcohol, cleared in Cedar Oil, and mounted in Balsam.

Koch-Ehrlich Method.

In Gentian-Violet Aniline-Water (p. 58), or Fuchsine Aniline-Water (p. 59), place sections or cover-glass preparations for at least 12 hours (more rapid staining of the latter is obtained by warming); then immerse in Nitric Acid (p. 62) for some seconds. Rinse in 60 p. c. Alcohol for some minutes, and then counterstain with Aqueous Solution of Vesuvine (p. 38) after Violet, and with Aqueous Solution of Methylene Blue (p. 36) after Fuchsine; rinse in water, dehydrate, clear, and mount in Balsam.

Nitric Acid is apt to injure delicate sections. To avoid this

MICRO-ORGANISMS. 69

Watson Cheyne recommended that the sections, after staining in the Fuchsine Aniline Water, should be transferred to Distilled Water, rinsed in Alcohol and placed in the following contrast stain for one or two hours:—Saturated Alcoholic Solution of Methylene Blue, 20 cc.; Distilled Water, 100 cc.; Formic Acid (sp. gr. 1·2), 1 cc.

Ziehl-Neelsen Method.

This method is applicable for staining the bacilli of tubercle and leprosy, and at the same time affords a diagnostic sign, for these are the only micro-organisms that can retain the stain after treatment with acid.

The sections are removed from weak Spirit into Neelsen's stain (Carbolic Fuchsine) (p. 59) for 10 or 15 minutes; then decolourised in Sulphuric or Nitric Acid (p. 62), rinsed in 60 p. c. Alcohol and washed in a large volume of Water, to remove the Acid.

Nitric Acid is apt to contain considerable traces of Nitrous Acid, which has a bleaching action on fuchsine-stained bacilli. Where this is suspected it is better to use Sulphanilic Nitric Acid (p. 62), the Sulphanilic Acid destroying any free Nitrous Acid that may be contained in the Nitric Acid.

For counterstaining, the sections may be treated as described for nuclear staining with Methyl Green or Methylene Blue (p. 36).

Dehydrate in Absolute Alcohol, clear in Cedar Oil, and mount in Balsam.

Kühne's Method.

Place sections for 10 minutes in Carbolic Fuchsine; thoroughly rinse in water; decolourise in Fluorescine Alcohol; then transfer to Ethereal Oil, Xylol and Balsam; or, to counterstain before mounting, sections may be transferred for 5 to 10 minutes into Methyl Green Aniline Oil, diluted with half its bulk of pure Aniline Oil; then pass into Ethereal Oil for 2 minutes, afterwards into Xylol, and mount in Balsam.

Gibbes' Method of Double Staining.

Cover-glass preparations are placed for 4 minutes in Gibbes' double stain, which has been previously slightly heated; sections should be placed in the stain at the ordinary temperature for some hours. The cover-glass preparations or sections are then washed in Methylated Spirit till no more colour comes away, dehydrated, cleared in Cedar Oil and mounted in Balsam.

LEPROSY BACILLUS.

The same staining methods may be employed as for the tubercle bacillus.

As leprosy bacilli stain much more quickly in the Fuchsine, the following plan has been suggested for distinguishing them:—

Sections are stained for 1 to 3 minutes in the Aqueous Solution of Fuchsine (p. 59); then placed for 30 seconds in a solution consisting of 90 p. c. Alcohol, 10 parts; Nitric Acid (Sp. g. 1·420), 1 part; dehydrate, clear, and mount in Balsam.

Treated in this way the leprosy bacilli appear well stained, while the tubercle bacilli are not stained at all.

ANTHRAX BACILLUS. (MILZBRAND BACILLUS.)

This may be stained with Violet by **Gram's method** (p. 64) or one of the modifications; with Fuchsine or Methylene Blue by **Kuhne's methods** (pp. 66 and 67).

GLANDERS BACILLUS. (ROTZBACILLUS. BACILLUS MALLEI.)

Cover-glass preparations are stained with Methylene Blue by **Kuhne's method** (p. 66) or by **Löffler's method** (p. 65).

Schutz's Method.

Sections (and cover-glass preparations) are stained for some hours in Aqueous Solution of Methylene Blue, differentiated in very weak Acetic Acid ($\frac{1}{2}$ p. c.) dehydrated in Alcohol, cleared in Cedar Oil, and mounted in Balsam.

SYPHILIS BACILLUS.
Lusgarten's Method.

Sections are placed in Gentian-Violet Aniline-Water (p. 58) for 12 to 24 hours at the ordinary temperature of the room, then for 2 hours at 40° C.

The sections are transferred to Absolute Alcohol for a few minutes, then placed for 10 seconds in 1·5 p. c. solution of Permanganate of Potassium, and washed in Sulphurous Acid. If the ground substance of the sections is not completely decolourised, the second part of the process must be repeated. After this the sections are dehydrated, cleared, and mounted in Balsam.

These bacilli, after staining by the Koch-Erhlich or Ziehl-Neelsen methods (unlike the tubercle bacilli), are easily decolourised by mineral acids.

Giacomi's Method.

Cover-glass preparations are stained for a few minutes in a hot solution of Fuchsine, then placed in Water containing a few drops of solution of Perchloride of Iron; then decolourised in strong solution of Perchloride of Iron. Sometimes a precipitate is formed with the Iron Solution, then the decolourisation is completed in Alcohol. Counterstain with Vesuvine.

PNEUMO-COCCUS.
Gram's Method

For staining the cocci.

Sections which have been deeply stained in Gentian-Violet

Aniline-Water, are rinsed in Alcohol; placed in Gram's Solution (p. 61) rinsed in Alcohol, cleared and mounted. The micrococci become a dense blue on a faintly yellow ground substance.

Friedlaender's Methods

For staining the capsule as well as the cocci.

Cover-glass Preparations.

Cover-glass preparations of the sputum are treated for 3 minutes with a 1 p. c. solution of Acetic Acid. The excess of Acid is then removed by filter-paper and the cover-glass is allowed to dry; the preparations are next placed in Gentian-Violet Aniline-Water (p. 58) for half a minute; then washed in Water, dried, and mounted in Balsam.

For Tissues the following Plan is used:—

Sections are placed for 24 hours in the following solution in a warm place:—Concentrated Alcoholic solution of Gentian Violet, 50 cc.; Distilled Water, 100 cc.; Glacial Acetic Acid, 10. cc. They are then placed from 1 to 2 minutes in 0·1 p. c. Acetic Acid; dehydrated, cleared, and mounted in Balsam.

BACILLUS OF DIPHTHERIA.

To stain these Löffler used his process, described p. 65.

They may also be stained by Gram's method (p. 64).

A piece of the false membrane is placed on a cover-glass and treated as a cover-glass preparation.

BACILLUS OF ENTERIC FEVER.

Gaffky's Method.

The organ from which the sections are made must be hardened in Alcohol.

The sections are placed for 20 to 24 hours in a deep blue opaque solution, which must be freshly made by adding a saturated Alcoholic solution of Methylene Blue to Distilled Water; they are then washed in Distilled Water (not acidulated), thoroughly dehydrated in Absolute Alcohol, cleared in Oil of Turpentine, and mounted in Balsam.

These may also be stained by **Löffler's Method** (p. 65).

It is usually stated that Gram's Method is not applicable as it removes the stain from the bacilli, but Woodhead affirms that if the sections containing these bacilli are allowed to remain for 10 minutes in a $\frac{1}{5}$ p. c. solution of corrosive sublimate, and then stained by Gram's method the bacilli are deeply coloured.

ACTINOMYCOSIS.

(i.) **Weigert's Method.**

Sections are immersed for 1 hour in Orseille (p. 60); they are quickly rinsed with Alcohol and counterstained with Gentian Violet.

If it is desired to stain the mycelium also, the sections, after being stained with Orseille, are submitted to Weigert's modification of Gram's method (p. 64).

(ii.) **Plaut's Method.**

Sections are placed for 10 minutes in Gibbes' Magenta solution (p. 59) (or Carbolic Fuchsine) at 45° C.; next they are rinsed in Water and placed in saturated aqueous solution of Picric Acid mixed with an equal volume of Absolute Alcohol for 5 to 10 minutes; they are then washed in Water, passed through 50 p. c. Alcohol into Absolute Alcohol, cleared in Cedar Oil, and mounted in Balsam.

(iii.) **Modification of Plaut's Method.**

Sections are placed for 10 minutes in Carbolic Fuchsine (p. 59) and decolourised for 24 hours in Fluorescine Alcohol.

Stain the nuclei with Ehrlich's Hæmatoxylin (p. 24) and counterstain faintly with Benzopurpurine (p. 40).

(iv.) **Babes' Method.**

Stain sections in Babes' Safranine solution (p. 60) for 2 minutes, rinse in Alcohol, and decolourise in Gram's solution (p. 61).

Only the clubs remain stained.

MAKING COVER-GLASS PREPARATIONS.

To thoroughly clean the cover-glass before use, wash it in Solution of Potash, rinse it in Diluted Nitric Acid, and afterwards in Alcohol.

In the case of blood, urine, and fluid discharges, a little is taken upon the point of a platinum needle which has been previously heated to redness, spread out upon the clean cover-glass, allowed to dry, and the cover-glass passed three times through the flame of a spirit-lamp to coagulate the Albumen.

In the case of sputum it should be rubbed to and fro over the cover-glass with a glass needle (thin glass rod drawn out to a point) until a uniformly thin layer of mucus is formed on the cover-glass. It is then allowed to dry; the drying can if necessary be accelerated by a gentle heat, and in the case of tough sputum a vertical stream of air from an air-ball syringe is useful. When dry the preparation is passed three times through the flame as above.

When examining **tubercular sputum,** the yellow caseous masses should be selected.

MAKING COVER-GLASS IMPRESSIONS.

A perfectly clean cover-glass is placed upon the culture, gently pressed, and then carefully raised. It is allowed to dry, passed three times through the flame of a spirit-lamp, stained by one of the methods, and mounted in Balsam.

STAINING COVER-GLASS PREPARATIONS.

For micro-organisms (when special methods are not employed) the following plan may be adopted:—2 or 3 drops of aqueous solution of either Fuchsine, Methylene Blue, Methyl Violet, or Gentian Violet, are placed on the cover-glass preparation; after a couple of minutes the superfluous stain is washed off with Distilled Water; the cover-glass can now be examined if it be placed, preparation downwards, on a glass slip, and the excess of Water removed with blotting-paper, or if a permanent preparation be desired the excess of Water is removed with blotting or filter paper, the cover-glass is then allowed to dry, and finally mounted in Balsam.

In place of the above aqueous solutions, Vesuvine dissolved in Glycerine (p. 57) may be used, the film is then washed and mounted in undiluted Glycerine.

Cover-glass preparations may also be treated by any of the methods described for use with sections.

STAINING TUBERCULAR SPUTUM.

A cover-glass preparation of sputum (p. 74) is placed in Carbolic Fuchsine (p. 59) warmed in a watchglass until steam rises for 3 minutes; it is then rinsed in Water to remove the adhering stain, decolourised in Nitric Acid (p. 62) (preferably Sulphanilic Nitric Acid), washed in 90 p. c. Alcohol to remove the Acid, then placed in aqueous solution of Methylene Blue (p. 36) for 1 minute, and well washed in Water to remove excess of stain; dried and mounted in Benzol Balsam.

If there be an objection to the use of Nitric Acid, then Sulphuric Acid and Alcohol (p. 62) may be used as a substitute.

STAINING SPORES.

If the cover-glass preparation be heated for much longer than in the ordinary method for bacilli, the spores will be stained as readily as the bacilli.

After a cover-glass preparation is made, it is passed several times (12) through the flame, or heated to a temperature of 210° F. for half an hour, or exposed to the action of strong Sulphuric Acid for a few seconds; then stained with aqueous solution of Fuchsine, Methylene Blue, or Gentian Violet.

DOUBLE-STAINING SPORE-BEARING BACILLI.

Neisser's Method.

Cover-glass preparations treated in the ordinary way (not specially heated as in the foregoing method) are immersed for 20 minutes in the Fuchsine Aniline-Water (p. 59) heated to 80° or 90° C. (176° or 194° F.); the cover-glass is then rinsed in Water, Alcohol, or weak Acid, according to the nature of the bacilli; counterstained with aqueous solution of Methylene Blue, rinsed in Water, dried, and mounted in Balsam.

The spores are stained red, and the other part of the bacilli blue.

STAINING FLAGELLA.

(i.) ### Koch's Method.

Cover-glass preparations are immersed in 1 p. c. aqueous solution of Hæmatoxylin; they are then transferred to a 5 p. c. solution of Chromic Acid, or to Müller's Fluid, dried and mounted in Balsam.

The flagella are stained a brownish black.

(ii.) **Crookshank's Method.**

Cover-glass preparations are stained with a drop of concentrated alcoholic solution of Gentian Violet; the cover-glass is then rinsed in Water, allowed to dry, and mounted in Balsam.

(iii.) **Löffler's Method.**

The stock solution is made by adding to 10 cc. of a 25 p. c. Aqueous Solution of Tannin, 5 cc. of cold saturated Aqueous Solution of Ferrous Sulphate, and 1 cc. of Aqueous or Alcoholic Solution of Fuchsine or Methyl Violet.

As the treatment, however, varies with the different micro-organisms, the reader is referred to an abstract of the paper in *Jour. R.M.S.*, 1890, p. 678.

DEHYDRATION.

As none of the usual clearing agents, used previous to mounting in Balsam, will mix with aqueous fluids, sections must be dehydrated before clearing. The usual dehydrating agent is Alcohol, and the only point which requires notice is the strength to be employed.

Preparatory to clearing in Oil of Cloves or Bergamot, 90 p. c. Alcohol answers every purpose. But for Oil of Cedar or the other members of the Xylol group, Alcohol of at least 96 p. c. must be used, and Absolute Alcohol is preferable.

Celloidin Sections.—Oil of Bergamot clears Celloidin without dissolving it, and as this oil can be used after 90 p. c. Alcohol, it possesses great advantages.

Cedar Oil also clears Celloidin, but as the film is quickly acted upon by Absolute (98 p. c.) Alcohol, it is safer to use a slightly hydrated Alcohol 96 p. c. (sp. gr. 0·807 at 60° F.), made by adding 2 cc. of Water to 120 cc. of Absolute Alcohol. This will clear into either Oil of Cedar or Xylol, and in 15 minutes (the maximum time required for dehydrating sections) will not perceptibly attack the celloidin. These sections can also be dehydrated in Absolute Alcohol, but it must be done quickly (about 2 minutes) or the celloidin film will be perceptibly affected.

The only other dehydrating agent that may be mentioned is Aniline, used in special circumstances where strong Alcohol must be avoided. This clears readily from 70 p. c. Alcohol, and is replaced before mounting by Cedar Oil, Benzol, or Xylol.

CLEARING AGENTS:

BEFORE MOUNTING IN BALSAM.

Those most generally recommended are Creosote, Turpentine, Xylol, and the Oils of Bergamot, Cedar, Cloves, and Origanum.

The object of using a clearing agent is to replace the Alcohol in the dehydrated section by a liquid which has a refractive index about the same as the Balsam into which it is to be placed and which will mix readily with it.

1. The fluid should clear quickly from a somewhat hydrated Alcohol.
2. It should not dissolve the aniline colours nor celloidin.

No one clearing agent is perfect in both respects, for it happens that the more readily the agent mixes with 90 p. c. Alcohol, the more liable it is to dissolve the aniline colours.

For clearing **celloidin sections**, Bergamot or Cedar Oil are the best. See Dehydration.

OIL OF BERGAMOT.

This oil will clear directly from 90 p. c. Alcohol. Its solvent action on Aniline Colours is very slight, except in the case of Methyl Green and the Violet dyes, which may be appreciably dissolved. Apart from this, its powerful odour and high price are the only objections to its general use.

It clears Celloidin without dissolving it.

CLOVE OIL.

This will mix in any proportion with 90 p. c. Alcohol without becoming turbid, and clears rapidly, but it dissolves the aniline colours to a very considerable extent.

The same objection applies even more forcibly to Wood-tar Creosote, and Carbolic Acid.

XYLOL.

This, on the other hand, is practically without action on aniline colours, but has so little affinity for water that it can only be used after Alcohol of at least 96 p. c.

Taking therefore Clove Oil and Xylol as types, the following have been classified as resembling the one or the other in reference to their behaviour with 90 p. c. Alcohol, and their action on aniline colours.

This strength of Alcohol has been chosen because it is that of the Methylated Spirit generally sold in this country and which for economy is largely used in washing and dehydrating the material for microscopic work. An agent, therefore, which will clear from this strength of Alcohol is a desideratum.

OIL OF TURPENTINE

has a comparatively low refractive index, resembles Xylol, but causes shrinkage and renders the tissues brittle. It becomes resinified on exposure to air.

TEREBENE.

It belongs to the Xylol group, but acts slightly upon aniline colours, and is open to the same objection as Oil of Turpentine regarding shrinkage.

BENZOL and TOLUOL.

These only differ from Xylol in their volatility; Benzol being the most volatile, and Xylol the least so of the three;

the rapidity of evaporation being in the ratio of 4, 5, and 9.

It is curious that, in two excellent books on this subject, Xylol is described as being more volatile than Benzol in one of them, and Toluol in the other.

Benzol has also been described as a solvent of the aniline dyes.

All three are practically without solvent action upon aniline colours.

OIL OF CEDAR WOOD.

Although an essential oil it resembles the type Xylol, but evaporates slowly and not to dryness.

It has very little solvent action on the aniline colours. It clears rapidly from Absolute Alcohol, but not well from 90 p. c. Alcohol. Sections can be left in it for several days without becoming brittle. It is a convenient medium in which to examine tissues before mounting them permanently.

It clears Celloidin without dissolving it.

OIL OF ORIGANUM.

This is Oil of White Thyme more or less adulterated; it dissolves the aniline colours to about the same extent as Oil of Cloves. Another oil, probably Oil of Marjoram, is sold as Oleum Origani Cretici, but this also dissolves the aniline colours.

BEFORE MOUNTING IN GLYCERINE.

Liquid Carbolic Acid, Liquor Potassæ, Alcoholic Solution Potash, Liquor Ammoniæ, Solution of Chloral Hydrate, Eau de Javelle, Eau de Labarraque, are all used for clearing vegetable sections for examination. Wash in Water, except after Alcoholic Potash when use 50 p.c. Alcohol, and mount in Glycerine or Glycerine Jelly.

ALCOHOLIC SOLUTION OF POTASH.

Hydrate of Potassium, 1 grm.; Distilled Water, 10 cc.; Rectified Spirit, 10 cc.

Dissolve the Potash in the Water and add the Spirit.

SOLUTION OF CHLORAL HYDRATE.

Chloral Hydrate, 30 grms.; Distilled Water, 10 grms.

SOLUTION OF POTASSIUM HYPOCHLORITE (EAU DE JAVELLE).

Chlorinated Lime, 20 grms.; Carbonate of Potassium, 20 grms.; Distilled Water, 200 cc. Triturate the Chlorinated Lime with half the Water; dissolve the Carbonate of Potassium in the other half; mix. After an hour, filter.

SOLUTION OF SODIUM HYPOCHLORITE (EAU DE LABARRAQUE).

It is made in the same way as above, substituting Crystallized Carbonate of Sodium, 40 grms., for the Carbonate of Potassium, 20 grms.

Both these Hypochlorite Solutions are also bleaching agents.

MOUNTING MEDIA.

When sections are mounted with Balsam, Dammar, Farrant, Glycerine and Gum, or Glycerine Jelly, it is not a necessity that the cover-glasses should be fixed with cement, as these fluids solidify at the edges, and so fix the cover-glass.

Some cement (Caoutchouc by preference) must be employed when Glycerine or aqueous solutions are to be used.

CANADA BALSAM.

A soft substance which is usually dried over a water-bath until, when cooled, it is brittle. It is then dissolved in Benzol or Xylol to form a somewhat thin solution; about 200 grms. of Balsam to 100 cc. of Benzol: the Xylol solution may be rather thicker, as it evaporates more slowly. Neither of them will affect the Aniline stains.

Mounts in **Benzol Balsam** harden in a day or two; those in **Xylol Balsam** remain soft for a week or two. I therefore much prefer to mount in Benzol Balsam.

Transfer the sections to a slide (using for thin sections a glass rod, p. 19), remove excess of clearing agent by gently pressing on the section with clean blotting-paper, add one or wo drops of Balsam Solution, and apply the cover-glass.

DAMMAR SOLUTION.

Gum Dammar dissolved in Benzol or Xylol.
About 100 grms. of Dammar to 100 cc. of Benzol.

DAMMAR AND MASTIC.

Gum Dammar, 100 grms.; Oil of Turpentine, 200 cc. Dissolve and filter. Gum Mastic, 50 grms.; Chloroform, 200 cc. Mix the two solutions.

This is an old formula which is still used by some workers.

GLYCERINE.

Two or three volumes of Glycerine to one of Distilled Water is a good medium in which to examine tissues transferred from aqueous fluids. For mounting permanently, the objects should be immersed in the above diluted Glycerine, which may be increased in strength if desirable; delicate objects may be placed in a mixture of Glycerine, 1; Alcohol, 1; Water, 1; which is then allowed to concentrate by evaporation.

GLYCERINE JELLY.

Immerse 100 grms. of French Gelatine in Chloroform Water, and, when soft, drain away the excess of water. Dissolve the softened Gelatine in 750 grammes of Glycerine over a water-bath, and add 400 grammes of Chloroform Water with which has been incorporated about 50 grammes of fresh egg albumen, and mix thoroughly. Heat the mixture to boiling for about five minutes to coagulate the albumen; make up the total weight to 1,550 grammes with Chloroform Water. Filter in a warm chamber, but protected as much as possible from evaporation.

Tissues can be mounted in this direct from water, but it is better to immerse them for a little while in equal parts of Glycerine and Distilled Water. Place them on a slide; remove superfluous fluid, and apply a few drops of the melted

Jelly; breathe on a cover-glass, place it on the fluid, apply gentle pressure, and set aside to cool.

FARRANT'S MEDIUM.

Dissolve 1 grm. of Arsenious Acid in 200 cc. of Distilled Water. In this fluid dissolve, at the ordinary temperature, 130 grms. of Gum Acacia, with frequent stirring; add 100 cc. of Glycerine; mix. Filter the solution through fine Swedish paper upon which has been deposited a thin layer of Talc.

This reagent is recommended because tissues mounted in it preserve their normal appearance, and it is more convenient than Glycerine in that it dries at the edges, and fixes the cover-glass.

Tissues may be mounted in this direct from Water, but it is better to soak them for a short time in dilute Glycerine before mounting.

Place the section on a slide, remove superfluous fluid, add 1 or 2 drops of the medium, and apply the cover-glass.

Formic Farrant.

Farrant's Medium to which 1 p.c. of Formic Acid (sp. g. 1·2) has been added.

GLYCERINE AND GUM.

Chloroform Water (1 in 200), 200 cc. Dissolve in this fluid 130 grms. of Gum Acacia, with frequent stirring, at the ordinary temperature of the air; add Glycerine, 100 cc.; mix. Filter as directed for Farrant.

This may be used in the place of Farrant's Medium when the presence of Arsenious Acid is objectionable.

IODINE MOUNTING FLUID.

Liquor Iodi B.P., 100 cc.; Water, 300 cc.; Glycerine, 200

cc. Mix, and dissolve in it 260 grms. Gum Acacia, with frequent stirring, at the ordinary temperature.

This is Farrant's Medium, in which Iodine is substituted for Arsenic. It is used for mounting tissues stained with Iodine.

ACETATE OF POTASSIUM.

Acetate of Potassium, 250 grms.; Water 100 cc.; dissolve by a gentle heat, and filter.

ACETATE OF COPPER.

Acetate of Copper, 1 grm.; Glacial Acetic Acid, 1 cc.; Camphor Water, 250 cc.; Glycerine, 250 cc.; Corrosive Sublimate, 4 grms.

For preserving and mounting Green Algæ. The Glycerine is sometimes omitted with advantage, as in the case of Volvox.

NORMAL OR INDIFFERENT FLUIDS.

These may be described as fluids which will not injuriously affect tissues. They consist of—

(i.) **BLOOD SERUM.**

Obtained by allowing blood to coagulate, and pouring off the serum after a day.

(ii.) **AQUEOUS HUMOUR.**

Best obtained by puncturing the cornea of a freshly excised ox's eye-ball.

(iii.) **NORMAL SALINE SOLUTION.** (0·75 p. c.)

Colourless Rock Salt or recrystallized Sodium Chloride, $7\frac{1}{2}$ grms.; Distilled Water, 1000 cc.

DISSOCIATING FLUIDS.

These fluids are used to soften the cement substance of connective tissue, epithelium, &c., in order that the various elements of the tissue may be separated and examined.

(i.) **IODISED SERUM.**

Solution of Iodine (B.P.), 1 cc.; Serous Fluid, 100 cc.

Should the brown colour fade, add a little more of the Solution of Iodine.

Place a piece of the tissue to be dissociated, which may be about $\frac{1}{10}$ in. diameter, in about 5 cc. of the fluid in a glass-stoppered bottle, taking care to maintain the brown colour of the fluid by adding a little of the Iodine Solution. After the tissue has been in this solution for a couple of days it may be teased out with needles.

This fluid is useful for macerating white nerve fibres.

(ii.) **DILUTE ALCOHOL,**

Alcohol (90 p. c.), 1 part; Distilled Water, 2 parts.

Takes about 24 hours to dissociate epithelium, connective tissue, &c.

(iii.) **BICHROMATE OF POTASSIUM,**

Bichromate of Potassium, 1 grm.; Water, 500 cc.

Used for epithelium and nerve cells of spinal cord. Takes from 1 to 2 weeks.

(iv.) CHROMIC ACID.

Chromic Acid, 1 grm.; Water, 1000—5000 cc.
For connective tissue, unstriated muscle, and nerve tissue.
Takes from 24 hours for nerve tissue, to 7 days for uterine fibroid.

(v.) NITRIC ACID.

20 p. c. aqueous solution.
Takes 24 hours to dissociate small pieces of muscular tissue; the muscle cells are then teased out, and may be mounted in Glycerine.

(vi.) NITRIC ACID AND GLYCERINE.

Glycerine, 1 part; Water, 3 parts; Nitric Acid, 1 part.
In dissociating nervous structure takes from 3 to 4 days.

(vii.) OSMIC ACID.

0·1 p. c. aqueous solution.
Used for cerebral cortex. Takes 24 to 48 hours.

(viii.) METHYL MIXTURE (Schiefferdecker's).

Methylic Alcohol, 5 cc.; Glycerine, 50 cc.; Distilled Water, 100 cc.
Used for retina and central nervous system and takes several days.

(ix.) CHLORIDE OF SODIUM.

10 p. c. aqueous solution.
Used for dissociating white fibrous tissue.

(x.) CAUSTIC POTASH.

30 to 50 p. c. aqueous solution.
It acts very rapidly.
Used for separating muscle cells, as in cases of myomata.

The tissues must be examined in this fluid, and cannot be preserved. If water is added to the dissociated tissues, they are soon dissolved.

(xi.) **SCHULZE'S MACERATING MIXTURE.**

Sections are placed in Nitric Acid sp. gr. 1·2, and about 2 or 3 per cent. of Chlorate of Potassium is added.

If the fluid be not heated, the sections should remain in it for several hours; but if the fluid be gently warmed until gas is given off freely, the result will be attained in a few seconds.

The tissue is washed in water, transferred to a slide, and the disintegration completed with needles.

Used in Botanical Histology for dissolving the middle lamella in vertical sections.

INDEX.

	PAGE
Acetate of Copper	10, 86
Acetate of Potassium	86
Acetic Acid and Alcohol	3
Acetine Blue	62
Acid Fuchsine	41, 48
Acid Magenta	42
Acid Rubin	41
Acidulated Alcohol	28
Acidulated Glycerine	28
Acidulated Water	61
Actinomycosis	73
Alcohol and Acetic Acid	3
Alcohol for Hardening	2
Altmann's Nitric Acid	8
Alum Carmine, Grenacher's	32
Alum Cochineal, Czoker's	32
Ammonia Carmine	30
Ammonia Carmine, Stronger	31
Ammonia Picro-Carmine	34
Amyloid	46
Aniline Black	48
Aniline Chloride	55
Aniline Nuclear Stains	35
Aniline Violets, General Methods	63
Aniline Water	61
Anthrax Bacillus	70
Aqueous Humour	87
Arsenic Acid	11
Babes' Method	74
Babes' Solution	60
Beale's Ammonia Carmine	31
Benda's Copper Hæmatoxylin	27
Benzol	80
Benzol Balsam	83
Benzopurpurine	40
Bergamot Oil	79
Bichromate of Ammonium	4
Bichromate of Potassium	4, 88
Bismarck Brown	38

	PAGE
Black Brown	60
Blood Serum	87
Blue Black	48
Böhmer's Hæmatoxylin	23
Borax Carmine	32
Canada Balsam	83
Carbolic Acid	80, 81
Carbolic Black Brown	6*
Carbolic Fuchsine	59
Carbolic Methylene Blue	5b
Carmine	28
Cedar Oil	81
Celloidin	15
Celloidin Sections	78, 79
Cellulose Staining	52
Chloral Hydrate	82
Chlorhydrin Blue	62
Chloride of Aniline	55
Chloride of Gold	43
Chloride of Sodium	87, 89
Chlor-Zinc Iodine	55
Chromate of Ammonium	4
Chromic Acid	6, 11, 89
Clearing Agents	79
Clove Oil	80
Congo Red	48
Corrosive Sublimate	7
Counterstaining Carmine	40
Counterstaining Hæmatoxylin	40
Cover Glass Preparations	74
Crookshank's Method	77
Czoker's Alum Cochineal	32
Dahlia Violet	38, 45
Dammar and Mastic	84
Dammar Solution	83
Decalcifying Fluids	11
Dehydration	78
Delafield's Hæmatoxylin	24
Diphtheria Bacillus	72
Dissociating Fluids	88

INDEX.

	PAGE
Eau de Javelle	82
Eau de Labarraque	82
Ebner's Solution	12
Ehrlich-Biondi Fluid	37
Ehrlich's Hæmatoxylin	24
Ehrlich's Hæmatoxylin Ammoniated	24
Ehrlich's Method	63
Elastic Fibres	45, 48
Embedding Media	14
Enteric Fever Bacillus	72
Eosine	40, 56
Erlicki's Fluid	6
Erythrosine	41
Farrant's Medium	85
Fatty Elements	43
Fixing Agents	7
Flagella Staining	76
Flemming's Solution	9
Fluorescine	62
Fol's Solution	9
Formic Farrant	85
Friedländer's Method	72
Fuchsine	39, 59, 67
Fuchsine Aniline-Water	59
Gaffky's Method	72
Gentian Violet	38, 58
Gentian-Violet Aniline-Water	58
Giacomi's Method	71
Gibbes' Double Stain	59
Gibbes' Magenta Solution	59
Gibbes' Method	70
Glanders Bacillus	70
Glycerine	84
Glycerine Gelatine	16
Glycerine and Gum	85
Glycerine Jelly	84
Gold Chloride	43
Golgi's Sublimate Method	51
Gram's Method	64, 71
Gram's Solution	61
Grenacher's Alcoholic Borax Carmine	32
Grenacher's Alum Carmine	32
Grenacher's Borax Carmine	32
Gum	14
Gum and Syrup	14
Hæmatoxylin	21
Hæmatoxylin, Ripening	23
Hamilton's Hæmatoxylin	24

	PAGE
Hardening Agents	1
Heidenhain's Hæmatoxylin	27
Hoffmann's Blue	39, 56
Hydrochloric Acid	12
Hydrochloric Acid and Glycerine	12
Hypochlorite of Potassium	82
Hypochlorite of Sodium	82
Intermediate Reagents	61
Iodine	46
Iodine Green	37
Iodine Mounting Fluid	85
Iodised Serum	88
Kleb's Glycerine Gelatine	16
Klein's Fluid	7
Kleinenberg's Hæmatoxylin	25
Kleinenberg's Picro-Sulphuric Acid	10
Koch-Ehrlich Method	68
Koch's Method	68, 76
Koch's Methylene Blue Solution	58
Kuhne's Methods	65, 66, 67, 69
Kuhne's Solutions	58, 59, 60
Leprosy Bacillus	70
Lignified Cellulose	53
Lithia Water	61
Lithium Carmine	33
Living Bacteria, Staining of	64
Löffler's Methods	65, 77
Löffler's Solution	58
Lusgarten's Method	71
Macerating Mixture, Schulze's	90
Magenta	39, 59
Magenta, Acid	42
Manipulation of Sections	18
Mastzellen	45
Methyl Green	36, 56
Methyl-Green Aniline-Oil	60
Methyl Mixture	89
Methyl Violet	46, 58
Methyl-Violet Aniline-Water	59
Methylene Blue	36, 57, 65
Methylene-Blue Aniline-Oil	58
Micro-Organisms	57
Mounting Media	83
Müller's Fluid	4
Müller and Spirit	5

INDEX. 93

	PAGE
Neisser's Method	76
Nerve Centres	29, 48
Nerve Endings	44
Nigrosine	48
Nitrate of Silver	44
Nitric Acid	8, 12, 62, 89
Nitric and Chromic Acid	13
Nitric Acid and Glycerine	89
Normal or Indifferent Fluids	87
Normal Saline Solution	87
Nuclear Staining	20
Oil of Bergamot	79
Oil of Cedar	81
Oil of Cloves	80
Oil of Origanum	81
Oil of Turpentine	80
Orange	41
Orseille	60
Orth's Lithium Carmine	33
Osmic Acid	8, 43, 89
Pal-Exner Method	50
Pal's Method	49
Pal's Solution	50
Palladium Chloride	29
Paraffin	16
Phloroglucin	55
Picric Acid	9, 13, 41, 56
Picro-Carmines	33
Picro-Lithium Carmine	35
Picro-Sulphuric Acid	10
Plasmatic Staining	39
Plaut's Method	73
Pneumo-Coccus	71
Potash, Alcoholic	82
Potash, Aqueous	81, 89
Preserving Sections	18
Rabl's Fluid	7
Ranvier's Lemonjuice Method	44
Renaut's Eosinated Hæmatoxylin	27
Renaut's Hæmatoxylin	26

	PAGE
Resin and Wax	18
Rose Bengale	47
Rubin, Acid	41
Rubin and Orange	42
Safranine	38, 47, 60
Safranine Aniline-Oil	60
Saturation Processes	15, 17
Saline Solution	87
Schiefferdecker's Mixture	89
Schulze's Solutions	55, 90
Schutz's Method	71
Sections, Manipulation of	18
Silver Nitrate	44
Soda Picro-Carmine	35
Specific Stains	43
Spores, Staining of	76
Sputum, Staining of	75
Staining in Bulk	28
Stains for Micro-Organisms	57
Sulphanilic and Nitric Acid	62
Sulphuric Acid Alcohol	62
Sulphuric Acid	62
Syphilis Bacillus	71
Terebene	80
Toluol	80
Tubercle Bacillus	68
Turpentine Oil	80
Vegetable tissues, see Cellulose	52
Vesuvine	38, 57
Victoria Blue	45, 59
Wedl's Solution	60
Weigert's Hæmatoxylin	26
Weigert's Method	48, 73
Xylol	80
Xylol Balsam	83
Ziehl-Neelsen Method	69
Ziehl-Neelsen Solution	59

www.ingramcontent.com/pod-product-compliance
Lightning Source LLC
Chambersburg PA
CBHW031119160426
43192CB00008B/1048